SECOND EDITION

STRATEGIC MANAGEMENT for the PLASTICS INDUSTRY

DEALING WITH GLOBALIZATION AND SUSTAINABILITY

SECOND EDITION

STRATEGIC MANAGEMENT for the PLASTICS INDUSTRY

DEALING WITH GLOBALIZATION
AND SUSTAINABILITY

Roger F. Jones

CRC Press
Taylor & Francis Group
Boca Raton London New York

CRC Press is an imprint of the
Taylor & Francis Group, an **informa** business

CRC Press
Taylor & Francis Group
6000 Broken Sound Parkway NW, Suite 300
Boca Raton, FL 33487-2742

First issued in paperback 2019

© 2014 by Taylor & Francis Group, LLC
CRC Press is an imprint of Taylor & Francis Group, an Informa business

No claim to original U.S. Government works

ISBN-13: 978-1-4665-0586-5 (hbk)
ISBN-13: 978-0-367-37966-7 (pbk)

Visit the Taylor & Francis Web site at
http://www.taylorandfrancis.com

and the CRC Press Web site at
http://www.crcpress.com

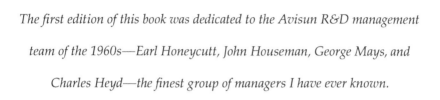

The first edition of this book was dedicated to the Avisun R&D management team of the 1960s—Earl Honeycutt, John Houseman, George Mays, and Charles Heyd—the finest group of managers I have ever known.

In this second edition, I am adding one more name—Steve Bowen, founder, president, and CEO of PlastiComp, the single most energetic and positive business chief executive I have ever had the pleasure of knowing.

Contents

Preface

This book is written for a broad audience, particularly aspiring professionals in the plastics industry who wish to become managers, as well as managers already in place who wish to round out their skills, consultants to the industry, university students, and faculty in plastics engineering and polymer chemistry departments. In the book, I use the term *manager*, rather than *executive*, because I believe the word is more inclusive. I define managers as including department heads as well as company officers (whom I would otherwise consider to be executives). Additionally, managers direct other managers and supervisors, while supervisors direct individual workers and professionals. The term *management* refers to this management group as directed by and including the senior executives. Most of the material presented here is oriented toward management, but some is also applicable to first-level supervision, though this is not the intended primary audience. A number of general management topics are discussed within the context of management in the plastics industry.

For purposes of this book, the term *plastics industry* is defined as referring to the development, manufacture, compounding, distribution, and processing/fabrication of plastics materials into products. Polymer processing machinery, additives, and other suppliers to the industry are mentioned in less detail due to the great variety of firms comprising this field and the fact that their involvement in plastics is often through divisions or business units of corporations whose main business is *not* plastics. This structure was a necessary compromise in order to keep the book from being overly broad, of reasonable length, and within the limits of my experience. The material presented is not only based on my own experience, but also on extensive research and interviews with managers throughout the industry.

There are some important changes from the first edition of this book to this new, second edition. First and foremost, the plastics industry is undergoing dramatic and painful changes due to the impact of increasingly globalized competition, as well as an unusually strong, simultaneous, and extended slowing of the world economy. Second, the industry is under increasing pressure from environmental groups and government regulators to improve its "sustainability." While both of these factors are speeding up the rate of change, they do not overturn the fundamental principles of how to manage successfully in the plastics industry. The first edition described the advent of e-commerce and the need to adapt to this innovation. E-commerce is old news now—readers with less than 10 years' experience in the industry will not recall a time without it—so a number of the more basic comments on this topic are omitted in the second edition.

In general, I have tried to describe "typical" situations in the various sectors of the industry, while commenting on the more interesting and important exceptions. The case histories are based on interviews with senior executives in the respective companies who were willing to be interviewed, and illustrate some examples of successful management in the industry.

I would be happy to hear from anyone who can add to the ideas set forth in this book.

Roger F. Jones
Broomall, PA

Acknowledgments—Second Edition

Both of this book's editions were the result of group efforts. While I am the sole author, I had a great deal of encouragement and assistance, without which I could have never finished this work or acquired the facts and stories necessary to inform readers about what it takes to run a successful business in the plastics industry. In the first edition, this included Peter Drucker of Claremont and New York universities, who passed away in the interim, but whose valuable work most assuredly lives on. In the second edition, I thank Allison Taub Shatkin of Taylor & Francis for persuading me to write this new edition, and my reviewers, Stan Verbraack, Paul Damm, Peter Lantos, Patrick Barron, and Ken Dargis, whose yeoman-like work was outstanding.

I also thank Bob Schulz (LNP, retired), Dave Hummel (Victrex), Troy Eubank (Modified Plastics), and Steve Maguire (Maguire Products) for sharing their thoughts with me about their businesses, as well as reviewing the summaries I wrote. Again, I thank my wife, Caryl, for her encouragement and support.

The Author

Roger Franklin Jones' 55-plus-years' career in the plastics industry has covered a broad range of technical and management functions as well as types of companies. He began with polymer producers, in technical positions in manufacturing and process development at DuPont (nylon intermediates) and ARCO Chemical (LDPE formulations), then product development and marketing at Avisun (polypropylene resin, film, and fiber), a joint venture of American Viscose and Sun Oil that was sold 7 years later to Amoco Chemicals (now part of Ineos). He then joined LNP Engineering Plastics, a small, rapidly growing, independent compounder, moving up through marketing and international operations management positions to COO, where he lead the rescue of the firm from imminent bankruptcy and its subsequent dramatic growth over the following 6 years, to become the largest independent proprietary compounder in the world. Beatrice Foods Co. acquired LNP in 1976, and he was then double-hatted as both LNP's president and a group executive in Beatrice's Chemicals Division with responsibility for two other companies, Dri-Print Foils (decorating foils) and Thoro System Products (specialty polymeric construction materials).

After 5 years in the Beatrice organization, he left to join a leveraged buyout consortium as managing partner, acquiring ailing Inolex Chemical Company, a manufacturer of plasticizers, urethane polyols, and cosmetics additives, from American Can Company; he was named Inolex's chairman and president. After restoring the company to profitability and growth, he sold his interests in Inolex, and soon afterwards was recruited by BASF as managing director of a newly formed engineering plastics business unit in North America. This was a successful grass-roots business start-up that included building acetal, nylon 6, and PBT production facilities, a compounding plant, and a technical service center. Following retirement from BASF, he founded Franklin Polymers, Inc., which distributed engineering/specialty plastics and offered industrial marketing/management consulting services. In 2000, he sold the distribution business and focused on consulting as president of Franklin International LLC. In 2004, he joined an investor consortium that founded PlastiComp LLC, a new technology company in the field of long-fiber thermoplastic composites, and was elected board chairman; he has continued to serve on the board as chairman emeritus since 2011.

He is a widely published authority on plastics and related topics. In the United States and overseas, he has authored over 100 articles and papers, and is inventor of record for 20 patents. In addition to the first edition of this book in 2002, he authored Hanser's *A Guide to Short Fiber Reinforced Plastics* (1998), and edited/contributed to two books published by the American Chemical Society, *The Chemical Industry and Globalization* (2006)—in its second printing—and *The Future of the Chemical Industry* (2009). His honors include the Honor Scroll of the American Institute of Chemists and election as a Fellow by both the Society of Plastics Engineers (975) and the American Chemical Society (2012). Within SPE, he has served in a number of section, division, and national positions, and has been a member/officer of the board of directors of SPE's Marketing and Management Division since 1990. He is a Life Fellow of the American Institute of Chemists (past offices include national secretary, national board vice chairman, Pennsylvania Institute president, Philadelphia Chapter chairman). He has been a guest lecturer at the universities of Delaware, Wisconsin, and Toronto, Winona State University, the Packaging Institute, and the Plastics Institutes of England and Australia.

Mr. Jones received a Bachelor of Science degree with honors in chemistry and honorable mention in English literature from Haverford College in 1952. At the graduate school level, he studied business administration at the University of Pennsylvania's Wharton School. He has completed language studies in German, French, Spanish, and Portuguese.

Soon after college graduation, he served as an officer in the United States Navy on active duty for 3 years at the end of the Korean War. He continued his military career in the Naval Reserve for an additional 30 years, retiring with the rank of captain. In the course of his naval service, he received two Navy Commendation Medals, a Letter of Commendation from the Secretary of the Navy, and a Meritorious Service Award from the commander, Naval Security Group. He was selected to command naval reserve units five times and served on admirals' staffs twice. He completed senior officer courses at the National War College, the Defense Intelligence School, and the National Security Agency.

Mr. Jones captained his college fencing team and was a member of US national teams sent to two World Fencing Championships. He was an alternate on the 1956 Olympic US Fencing Team. He won a number of collegiate and US amateur fencing titles, and served as chairman of the Philadelphia and Western New York divisions, vice president of the US Fencing Association and chairman of the National Rules Committee. For many years he was tournament director of the Middle Atlantic Collegiate Fencing Association. In 2011, he was elected to Haverford College's Thomas Glasser Hall of Achievement, the only fencer to have been so honored.

He is married to Caryl Jeanne Reisgen Jones; they will soon celebrate their diamond anniversary. They have three children and seven grandchildren. His father, Franklin D. Jones, ScD (Hon), was a chemical engineer who pioneered the discovery and development of plant hormones in the 1930s and 1940s.

Suggested Reading

A number of the following works were written quite some time ago, but the details and realities expressed in them are pertinent today as when they were written. The books written since 2002 have more detailed information on events of the last several years, particularly those relating to globalization and sustainability, which are applicable to plastics industry management.

Axelrod, A. *Patton on Leadership, Strategic Lessons on Corporate Warfare,* Prentice Hall 1999.

Drucker, P. *Managing for Results,* Harper & Row 1964.

Drucker, P. *The Effective Executive,* Harper & Row 1966.

Drucker, P. *Innovation and Entrepreneurship,* Harper & Row 1985.

Drucker, P. *The Essential Drucker,* HarperCollins 2001.

Jones, R. "US Independent Compounding—Past, Present, and Future," *Plastics Engineering,* May 1996.

Jones, R. "New Routes to Market in the 21st Century," *Plastics Engineering,* August 2000.

Jones, R. *The Chemical Industry and Globalization,* American Chemical Society, 2006.

Jones, R. *The Future of the Chemical Industry,* American Chemical Society, 2009.

Peters, T. and Waterman, R. *In Search of Excellence,* Warner Books, 1982.

Tolinski, M. *Plastics and Sustainability,* Scrivener Publishing, 2012.

1

Introduction

1.1 Why a Management Book for the Plastics Industry

There are a great many excellent books on industrial management written from a very broad standpoint but none appear to deal with the specific conditions of the plastics industry. The plastics industry has become a major part of the world economy during the last half of the 20th century. Although extensive industry restructuring taking place since the 1990s has led some observers to believe that plastics is now a mature business sector, this is a mischaracterization. No industry can be properly called mature that typically grows at multiples of the gross domestic product (GDP) and finds new uses virtually every day. While the plastics industry is being affected by the ongoing globalization of competition, not to mention the Great Recession that began in 2008, these conditions affect all types of manufacturing in virtually all countries worldwide. Nevertheless, the plastics industry overall has long ago gone beyond being a specialty business, and a number of segments have indeed become commoditized. Besides globalization, restructuring is being driven by the transition of a number of former specialty segments into semi-commodities. Management of each of these segments and the transitions between them present a number of challenges that differ significantly, as well as differing from those found in truly mature materials industries that grow at the GDP rate or less. This book tries to highlight these differences and show how to deal with them effectively. Other plastics industry management issues that diverge from more general treatments of these topics include key factors that differ for individual industry segments, the way product and process technology define the business that one is in, staffing, and the effective use of patents and trade secrets. Some more general management issues are also included to present the plastics concerns in a seamless matrix, as well as to indicate the author's point of reference.

Management is as much an art as it is a science. Although one can and does measure just how successful the management of an enterprise has been via financial analysis, the foundational building blocks of the management process that produces these results are *human relationships*, which cannot be reliably quantified. Even so, there are a number of management principles that can be applied with a reasonable expectation of results. One may discover these principles and when to apply them through trial and error, or learn from the experience and insight of others. The book will endeavor to explain which management techniques generally work and which do not, based primarily on the author's observations and experience in the plastics industry, but also on those of others whom the author knows personally and respects. Most of these techniques are essentially timeless even if, from time to time, they appear to have been temporarily eclipsed by others. The emphasis is on the practical and the applied, rather than the theoretical.

Benjamin Franklin wrote in *Poor Richard's Almanac*, "Experience keeps a dear school, but fools will learn in no other." To add to that thought, the most expensive mistakes are those made by senior executives. This book will try to point out how to avoid making more egregious errors without becoming paranoid about making mistakes. It's surprising but readily observed that some specific errors seem to be repeated over and over again in the plastics industry, mainly in the areas of acquisitions, but also in transitions from one type or size of business to another. It would appear that most of these seeming oversights stem either from ignorance or from oversized management ego. The most common or outstanding lapses will be analyzed in sufficient detail that you, the reader, can have the benefit of someone else's tuition bill. However, this exercise is not conducted for the purpose of holding anyone up to ridicule, because everyone makes mistakes in life. The author's concern is that you *learn* from your mistakes as well as those made by others, and not repeat those mistakes blindly. In the familiar words of George Santayana, "Those who cannot remember the past are condemned to repeat it."

The plastics industry is founded on the bedrock of science and engineering. Those who work in this industry are, by and large, scientists and engineers who have learned the enormous value of the scientific method and to apply it to all aspects of their work. The scientific method calls for the thorough testing of a hypothesis both to prove *and* to disprove it, before communicating the findings to colleagues for comment and criticism. Indeed, a hypothesis cannot be considered proven until *other* scientists and engineers have been able to duplicate those same results through independent testing. The objective, in all cases, is both to establish an explanation of a finding and also the boundaries that limit the understanding of those findings. The scientific method can and should be applied in management wherever feasible, recognizing, of course, that the human factor introduces variables that cannot be controlled. Therefore, results in human relations may be reproducible, say, only 85 times in 100 tries, not the usual standard of 99 out of 100. It is critical to distinguish between assumptions based on anecdotal data and the results of scientifically designed experiments, to ensure that controls have been used, the number of data points is statistically meaningful, and that the results can be duplicated. This approach applies to lessons learned from experience, most certainly. Anecdotal experience can be very misleading and needs to be verified insofar as possible. Computer models are emphatically *not* the same thing as actual experimentation. Too often, faddish management or personnel policies have been adopted because a single or very few prior uses of it appear to have yielded positive results. Wishful thinking is no substitute for the scientific method, under any circumstances. The author has made it a point to apply these principles insofar as possible before recommending management tools based on his own experience and that of others.

These points are emphasized strongly because one cannot help but observe so many instances where they are ignored in the industry, to its detriment.

1.2 Management as a Career

A professional engineer or scientist should be certain that he or she really wants to become a manager before taking the plunge; doing so means quite a change in one's work life. First of all, unless you really enjoy working with other people, don't even think about a career in management. *All* of your results will be accomplished by others, whom you must train, motivate, and evaluate. If this is unappealing, then you will neither enjoy being a manager nor be an effective one. Managers must delegate tasks to others to accomplish rather than carrying out those tasks themselves. This also frequently means learning to live with work done to less perfect standards than if one had done the work personally.

Second, being a manager means a major shift in the nature of one's work. Most professionals take satisfaction in seeing a number of individual projects through to completion, whereas a manager's job is continuous for the most part, with few defined starts and finishes other than those set by the arbitrary dates of a fiscal year.

Third, being a manager will demand a personal commitment of much more than 40 hours per week, especially in start-up or work-out (on the verge of bankruptcy) situations. However, under normal conditions, it has been the author's experience, as well as that of many others, that something is wrong with the approach of managers who consistently work more than 60 to 80 hours per week, or fail to take regular vacations. Their problem likely results from one or more of the following reasons:

- Doing their subordinates' tasks for them (micromanaging)
- Immersing themselves so deeply in details that they have trouble in seeing the overall picture of their company and the future direction charted
- Failing to prioritize their objectives by making every task of equal importance

- Failing to identify and limit their list of personal objectives to those that are critical to success and can be done only by the manager
- Seeking out and accommodating every point of view or splitting the difference between them, rather than deciding on a single course of action and carrying it out
- Are not competent to handle the work
- Any combination of or all of the above

A manager also needs to maintain a healthy family life, as well as make time for community involvement. For most people, their family is the most important focus of their lives. It's virtually a cliché that someone on their deathbed is unlikely to bemoan that they had not spent enough time in their office! As in almost everything in life, moderation and balance are the keys to success.

Community involvement has at least two dimensions. The first is personal and what most people think of: religious, charitable, and other service-oriented activities. This is something in which everyone should participate as a responsible member of his or her community. It is part of the necessary balance in life just mentioned.

The second dimension is business related and most certainly not to be taken as a casual add-on: community liaison. Companies in the chemicals and plastics industries are under constant fire from environmentalist and other activist groups, many of whom are simply anti-industrial business. It is essential that management be a positive, visible factor in community relations and the concerns of its citizens. Remember that a number of your employees are also likely to be members of the community. You owe them the opportunity to feel proud of where they work and what they do. A proactive approach to community relations will establish a reservoir of credibility and goodwill that will help cope with activists' attacks over the issue *du jour*. Most importantly, it is the *right* thing to do; the surrounding residential community *should* know if there are any hazards to their well-being that could result from an accident or improper operation at your plant, *how* you will handle such a situation, and have a firsthand opportunity to assess your credibility for assurance that you *will* take the proper steps immediately under such circumstances. Needless to say, you must ensure that the means exist to deal with emergencies, and that they will be utilized, effectively and immediately. In the minds of members of the communities, emergencies can include the emission of unpleasant (they don't even have to be toxic!) odors. More than one company has tried to pass off an occasional stink as just something that should be accepted as the price of living near a plastics plant, and then were set upon by regulatory authorities and attacked in lawsuits as a result of management's attitude, which can be often perceived as cavalier and arrogant. On the other hand, if activists employ bad science and attempt to play on the fears and ignorance of the community about what the company does and its potential for endangerment of the community, management must be willing to stand up and objectively rebut any misinformation, point by point, without responding to personal attacks in kind. The community may yet believe those attacks if management has not previously demonstrated and communicated the nature of its business and the precautions taken, for example, full compliance with all applicable government regulations, Responsible Care standards, ISO 14000 environmental standards, functioning community liaison committees, etc., to assure safe and community-friendly operations. Losing money is not the only way to be put out of business.

When the author first moved into senior management, his company was in financial crisis, and a major change in company direction had to be made immediately or the company would fail. A number of managers get their start under somewhat similar circumstances—someone else has made a mess, and it's yours to clean up, *now*! You may or may not get a lot of help and be offered a bewildering (and likely conflicting) array of solutions, but ultimately it will be up to you and you alone to decide how to solve the problem. If you succeed, the credit will, and should be, shared between you and your team. If you fail, be prepared to accept the major share of the blame alone. This may not seem completely fair, but it's the nature of being a senior manager and you had best be prepared to accept such judgments if being a manager is your chosen career.

It's also wise to view a career in management as a series of stepping-stones. No one should ever contemplate that his or her current position in management or with a company will last a lifetime. Not only are the days long gone of a lifetime career at one company, but there are other good reasons, too. Professor William Meldrum, the author's college chemistry advisor, once told him that "a good chemist changes fields every ten years." The author has found that to be a maxim of great value in the course of his career, as well as from observing the careers of others. After ten years in a particular discipline or position, learning genuinely new things becomes increasingly infrequent, as does making more and greater contributions—when one becomes stale, it's time to move on. Change refreshes, purges, and renews those who embrace it. It sweeps away those who resist it. Always seek out new and greater challenges to meet, no matter what your age or status.

A successful career as a manager can and should bring financial rewards, but it will be a glass half empty if you do not find more than money for satisfaction. You can find great satisfaction in bringing a team together, challenging it to achieve high goals, and seeing it reach them. The author's greatest sense of accomplishment as a manager has come far more from helping many different people to succeed and find fulfillment in their jobs while creating thousands of satisfied users of my company's products and services, than from any financial rewards (although, to be sure, he never turned any of these down!). While one presumably could do this in any industry, the plastics industry has such a broad and diversified involvement in the economy that it would be truly exceptional if anyone found that they were bored by doing the same old thing, day after day. There is very little in the world of plastics that is *not* new and exciting, all the time.

Being a CEO, however, should never turn into an ego trip. CEOs who make themselves the story of a company are dangerous to the well-being of the company. Did such a CEO buy a company jet in which to travel, even if the company's sites are in locations served by scheduled airlines? Did such a CEO install an opulent office with costly artwork? These are signs that a CEO's primary interests have diverged from those of the company, its owners, and employees. Beware when the boss's face appears on the cover of one or more business magazines—the team approach has been lost when the boss is taking credit for what the team has done. The best CEOs are not interested in promoting themselves but in promoting their company and their team. The best CEOs don't spend money on their own gratification but on what helps the company and the team succeed. These types of CEOs usually have a pattern of using people, in the unpleasant sense of the word. If you find yourself working for one, you had better keep your resume current—you will need it sooner rather than later.

Many plastics companies are small, entrepreneurial firms where the founder hopes to see members of his or her family work for the firm and, in time, possibly manage it. This is a natural ambition and the children of such founders have a potentially wonderful opportunity presented to them. However—and this is a big qualification!—the emotional fit among parent and children must be such that all will be comfortable with working with one other. Will the siblings get along or will there be resentment if the family talent turns out not to have been spread equally? How much independence is the parent willing to allow the children to make their own decisions? Will other employees in the firm view the family as blocking their own ambitions? It has been the author's observations that these problems are greatly exacerbated if the children go right to work for Dad or Mom directly from school. The best way to reduce these natural frictions is for the children to go to work for another firm, where they can gain experience away from the parent and develop self-confidence in the process. It is difficult for children to mature and acquire a sound sense of their own self-worth and competence without some career experience in the world apart from their parents. Once they have this—and it could take a period of perhaps five to ten years—they ought to be able to move into the family business and begin making a contribution right from the start. The other employees, as well as the parent, will respect them more for having "earned their spurs" elsewhere first.

There is another myth about successful managers that deserves mentioning, if for no other reason than to refute it. It may be best known from Leo Durocher's famous line that "nice guys finish last." This is really just a variation of the myth that managers get more results through fear and intimidation than by being "nice." Frankly speaking, this is nonsense. While it is true that fear and intimidation will work—for a short while—it is also true that both subordinates and managers will burn out quickly in such a work environment. This philosophy might be a holdover from a medieval army command mentality that forcing the troops to storm the battlements was best achieved by making the foot soldiers understand that their chances of survival, however slim, would be better by attacking the enemy than being shot or stabbed from behind by their own officers. The record shows that there are plenty of nice guys who finish first. The usual mixture of human personalities found in management positions ensures that both kinds will be present. Managers who cannot focus on the long term are not thinking of the best interests of their companies, their stockholders, employees, customers, or suppliers.

1.3 What Six Things Management Must Do

There are a great many opinions offered at business schools and by industry executives about the proper functions of management. The author's observation on this critical subject is that there are six primary responsibilities that management must execute for a business to succeed:

- First, management must organize the business to meet market and customer needs.
- Second, management must *recognize and manage change.*
- Third, management must develop company goals and get everyone onboard with the plan.
- Fourth, management must continuously appraise subordinate performance and provide positive feedback while *not* micromanaging these same personnel.
- Fifth, management must *lead by example* while demonstrating the highest levels of honesty and integrity.
- Sixth, management must *ensure that the business is increasingly profitable.* This means taking the necessary steps to be certain that sales are made at profitable prices, new products are always under development, customers served, costs controlled, and all assets fully and gainfully employed. While this may sound laughably obvious, it is absolutely astonishing how many businesses fail because management allows itself to be distracted by other considerations, such as increasing sales volume without regard to profitability, or being a technology "pioneer" regardless of cost, or building an overly large staff during upswings in the business cycle.

Let's examine these guiding principles in more detail.

1.3.1 Organize the Business to Meet Market and Customer Needs

No business can exist without meeting market and customer needs. Even a genuine monopoly (which, as Peter Drucker says, "is as mythical a beast as a unicorn, save for politically enforced, that is governmental monopolies") would sooner or later find its offerings supplanted by cheaper, more effective alternatives from others. There are some managers whose idea of heaven on earth would be to sell out their plant's capacity to a single customer and then play golf for the rest of the year (this anecdote is from the direct experience of the author!). Sorry, but that's one dream that will never come true—and if it did, it would likely be followed by hell on earth as soon as the customer's business declined or a competitor took away the business, or a myriad of other things happened, all just because the supplier didn't want to deal with reality. What is that reality? It is that a good manager must be a bit paranoid, for all of those reasons just cited above. A manager would be grossly derelict if he or she permitted the company's business to depend on a single customer, end use, or market segment.

The way to avoid these problems is to organize your business to deliver what the customer *needs* (not just *wants*), where and when the customer needs it, at a cost that either permits better than competitive pricing or offers benefits that are worth more than competing products. The business must also be organized to replace customers who fall by the wayside and to gain new ones in the same, related, or new markets. The company must be organized so that its various functions work together to learn what customers want, make or buy those items, deliver them on a timely basis, and have money left over after collecting and paying bills. The company must be staffed by competent, motivated professionals who operate as a team, that are led—not bossed—by management to be customer-focused.

The worst error management can make is to become so engrossed in the *process* of managing that it mistakes the process itself for *results*. Results are achieved *only* when customers buy your products at a price that makes a profit for your firm.

1.3.2 Recognize and Manage Change

No business ever operated without encountering change. No business is ever protected from change. Change takes many forms. It can be internal, such as a transition from an entrepreneurial business culture to a managerial one, growth or contraction, the impact of an acquisition or divestiture, to name but a few. It can also be external, such as the emergence or disappearance of competition, certain markets or customers, new technologies, government regulation, etc. Management must be ever alert to recognize and adapt the enterprise to such change. This involves being both proactive and reactive, depending on the situation.

What are the warning signs of change? Some are obvious. As mentioned above, the emergence or disappearance of competition, markets, technologies, etc. These are not hard to notice, but they do require investigation to analyze how and why the changes took place. One should never operate on the basis of assumptions, because important changes in the direction your business is headed may be missed if the changes are dismissed as accidents or of no consequence for your business.

- Did a new competitor come into being because someone has discovered a new technology or are you not covering the market adequately?
- Did a competitor disappear because it was undercapitalized or the market itself is shrinking?
- How will a new technology affect your business and why didn't your company come up with it first?
- Are new environmental regulations the result of something stemming from industry-wide problems, your problems, or weak community relations on your company's part?
- Has your customer stopped using your product because they are buying from a competitor, have designed your product out of their own, cannot compete against other firms, or is the end-use market your customer supplies in decline?
- Have new customers or markets appeared because you were lucky or because you worked to develop them, and are you prepared to supply them?

Many forms of change are gradual and therefore not obvious. These are usually internal, such as the effects of growth (or the lack of it) on the company culture. One should be regularly looking for telltale signs that they are reaching the point that they require action. In the case of growth effects, the signs can include declining sales, excessive late product shipments, low employee morale and high turnover, quality problems, and infighting and turf wars between company departments. Management must deal promptly with these problems before they damage the company, attacking underlying causes as well as dealing with the symptoms. Often, reorganization is called for, and possibly a redefinition of the company's mission. All of these situations also involve the four other responsibilities of management.

Management must be alert to distinguish between genuine changes in the business and those that *appear* to be happening because "everyone knows" that they are taking place. It is not unique in the long history of business to find people swept up by fancied and imagined changes that required specified actions, and this problem is still very much alive and well today. When all about you are restructuring, it takes courage to recognize and state boldly that *your company may not need to do so*. Don't get caught up in fads. By the time a technique becomes a fad, its principal usefulness (usually "shock value") is likely to have passed. People in business cannot afford to ignore reality in favor of their fantasies for long. Reality seems to have a way of catching up much more quickly with businesses than it does with many other endeavors.

1.3.3 Develop Company Goals and Get Everyone on Board with the Plan

Developing company goals and a business plan is a critically important exercise in leadership. Top management must determine just what business the company is in—this task cannot be delegated. The definition of the company's business then becomes the target of the company's goals, but goals do not have to be defined solely by top management. On the contrary, it is critical to success to get the input of subordinates and to use this input wherever one can, when developing goals. It is the responsibility of management to exercise its judgment as to how much of this input to use, not just to put together an anthology of every idea submitted by subordinates. Management must remember that it has the final responsibility for goal setting. Why? Because it is impossible to accommodate every idea generated by your subordinates in a single set of simple, practical goals that are critical to the company's growth and health. Companies are financial organizations, not social ones, and they are certainly not democracies. H.B. Swope said it best: "I cannot give you the formula for success, but I can give you the formula for failure, which is: try to please everybody."

Goals are the "what" and business plans are the "how." The input from subordinates should be much more substantive in developing plans than goals. Good management delegates authority to execute plans to the lowest appropriate level, and it is appropriate that the people who will do this are the ones who have the most input about how it will be accomplished.

Subordinates' own goal setting and planning is often best when done from the bottom up within the framework established by management, including "stretch goals." When people have had some say in the development of their goals and plans, they will usually accept them far more readily than if they were imposed from above. When someone cannot or will not accept reasonable goals and plans, then it is time they moved on to another company. Life is too short for anyone to continue to work in a place where they are unhappy—and creating dissension. That goes for management, too. If you are fundamentally unhappy with your own situation and cannot find a way to alter it, then you cannot do your job properly and need to make a career change.

Remember that most people respond to what they perceive to be their own self-interest. There is nothing immoral or shameful in this; it is simply a normal fact of life. If people do not look after their own interests, it is unlikely that anyone else will. You must show each of them that working as a member of a team toward common goals is indeed in their best self-interest.

1.3.4 Continuously Appraise Performance and Provide Feedback

A leading cause of many small business failures is the inability of the boss to delegate authority to subordinates, that is, micromanagement. It goes on in larger businesses, too, but the sheer inertia of bigger firms makes it easier for this defect to be masked for a while before it has overtly damaged the business. Micromanagement is usually a sign that a manager is in over his or her head or has such a strong compulsion to control every aspect of a job that they are unsuited to the position they are in. A manager is paid to *supervise* subordinates' work, never to *do it for them*.

Also, please note that one can delegate *authority*, but not *responsibility*. That's because the boss is ultimately responsible for everything below his level and within his department in an organization. If something goes wrong, a manager should never attempt to hide behind a subordinate, but should accept responsibility for the error and for fixing it properly.

How does a manager delegate authority but not responsibility? By keeping informed as to how the subordinates are carrying out their duties and what results they are obtaining. The least intrusive way is via weekly verbal reports; a well-run meeting that lasts no more than an hour is a good way to supplement regular informal brief conversations, plus monthly written reports. An electronic office database can be useful in keeping track of what is happening within the company. Help your subordinates to develop a sense of when policy is involved and to come to you for guidance, but otherwise to handle matters on their own while keeping you informed. Learn to *coach* your subordinates, not to bark out orders. Direct orders are necessary on a battlefield when lives depend on immediate, unquestioning obedience, but they are appropriate only under rare circumstances in a business setting.

Positive feedback is an essential part of the process. Make sure you tell subordinates when they're doing things right—use a "did this well, do this differently" approach when giving feedback. Do this whenever a subordinate gives you a report on something substantive that they did. Don't let a subordinate keep making errors without sitting down with him or her to identify what the reason is and then developing an action plan to fix it. This is more than a matter of simple fairness; it's essential for effective personnel utilization. If the individual cannot make the necessary adjustments to meet his or her assigned goals and the company standards for the job, then a clear, written plan must be put in place, with the employee's participation, that identifies not only what has to be done and when, but also makes it clear that failure to execute on a timely basis will lead to employment termination. There is absolutely no excuse for employees with years of satisfactory performance reviews in their records suddenly being recommended for termination. When this happens, there is a serious performance problem on the part of *management*.

1.3.5 Lead by Example

One thing the United States military teaches that is also true in business is that the boss must lead by example. The boss can never hide behind a "do as I say, not do as I do" philosophy. Any boss who tries this will immediately and irretrievably lose all credibility and respect among his or her subordinates. As described earlier, the boss must always accept responsibility for the actions of subordinates—this is an act of loyalty that should and will earn the respect and loyalty of every subordinate worth their salt. Respect cannot be commanded or "deserved." It can only be earned. Furthermore, a manager must back up his or her own management, too. If a manager is disloyal to *his or her* own boss, it won't be long before that manager's subordinates lose *their* loyalty to him or her. If you don't trust your boss, why on earth are you still working for him or her?

Why this emphasis on loyalty? Because *trust* is based on loyalty. Any organization that lacks trust is doomed to failure, because no one in it can ever be sure what the motivations of others are or the accuracy of the information they are receiving. Loyalty is insufficient by itself, however. Managers have to demonstrate every day that their lives are ordered by a strong sense of right and wrong, that their personal integrity cannot be compromised, and that they are invariably honest in their dealings with everyone with whom they come in contact. Management cannot run a sound business by conducting affairs to stay just within the law or by getting the better of everyone else without exception. The need for loyalty and trust demands that you remove those people from your organization who have shown that they cannot be trusted. Furthermore, managers must also demonstrate consistent professional competency—people easily see through bluster masquerading as knowledge.

Some readers will say to themselves, "that sounds like a level of perfection that doesn't exist in the real world." Not true. You wouldn't accept an occasional shortage in your paycheck would you? Subordinates are entitled to expect the same kind of consistency. Yes, there will be some bad apples in every barrel, but they cannot and should not be used to define all the rest of the apples. One of the most transcendent elements of human society is to aspire to ideals that may not be attainable 100% of the time but nonetheless are very much worth striving for 100% of the time.

1.3.6 Ensure That the Business Is Increasingly Profitable

Don't ever apologize for making a profit in your business and wanting to make a bigger one—that's your job. Even Samuel Gompers, the 19th century American labor leader, said that the worst thing that could happen to an American worker would be for the company that employed him to lose money. One of the more annoying things about socialists and other anticapitalists is their constant characterization of "profit" as something evil. They depict private business ownership as akin to organized crime, as though employees, suppliers, and customers have been forced to do business with companies at the point of a gun. These critics never describe profits as "normal" or "acceptable," but always as "obscene" or "excessive." One is led to believe that they believe businesses should be run at breakeven (which is like balancing on the edge of a knife) or perhaps better, at a loss? As the wreckage of the former Soviet Union and Eastern European countries testifies, the idea that businesses should not be concerned about profits and losses cannot work for any length of time without a great deal of economic and social damage. Sooner rather than later, payrolls and creditors' bills have to be met.

In a free society, profitable companies pay their employees well, sustain a number of vendors, satisfy a number of customers, fulfill stockholders' investment objectives, and pay taxes to local, state, and federal governments (actually companies *collect* taxes for governments, as only *people* actually pay taxes). Companies that are unprofitable and cannot meet their bills are eventually sold or liquidated. In any event, a great many people lose their jobs beside the manager, and the stockholders' investment suffers or is lost altogether. This is the ultimate test of management. If you can't run your company to make money, you won't get to run your company for very long.

Are there no exceptions to this rule? Remember the Internet bubble of the 1990s—companies seemingly worth huge premiums in the stock market as long as their sales continually rose, even if they lost money—as in "cash burn?" It didn't take all that long for investors in those companies to find out that when the promise of earnings from those rising sales revenues were not forthcoming, their stock holdings crashed, and the companies were acquired for pennies on the dollar or liquidated. The long-term value in any new concept can look very appealing as long as the actual results are veiled in the mists of the future. However, the plastics industry is *not* the latest financial fad! The stock market, banks, and venture capitalists hold our industry to a demanding standard: it is absolutely unacceptable to lose money for any length of time. On the contrary, companies in our industry are expected to show *rising* earnings and improving returns on investment, more or less continuously.

One of the reasons that chemical and plastics companies' stocks do not command higher levels of valuation in the stock market is that these industries have had a long, sad history of boom-and-bust cycles. Today's management always has the opportunity to break with this well-earned stereotype, if it chooses to do so. It does not pay to continue to build capacity if each incremental sale shows a lower profit than the preceding one. In today's globalized marketplace, management must be very cautious about adding capacity, particularly with respect to the location of that capacity, if the product line shows any tendency toward becoming mature.

Ever hear the old joke, "we lose money on every unit sold but make it up on volume?" It never ceases to amaze how many managers think they can sell products at less than full cost as long as they recover out-of-pocket costs and "make a contribution to overhead." In the oil industry, this is called "selling incremental barrels." The big hole in this idea is that all those customers who are paying full price are bound to discover eventually that they too can buy the same product for less. Soon the company will find that it is now selling *mostly* incremental barrels instead of standard-ones and losing money on *most* of its sales. Price alone is always a miserably unimaginative choice for an inducement to place an order. Management must ensure that any program to increase sales does not rely on offering the lowest price, because it runs a strong risk of violating the basic rule of not making a loss. Successful selling depends on bringing *value* to the customer that goes beyond price. This and the earlier comments in this vein will be dealt with in more detail later in this book.

If your company loses money in a month, and there is no foreseen reason for it, you should establish the cause and correct the situation as soon as possible. If your company loses money for a quarter, then you had better rediagnose the problem and fix it *immediately*, because you are unlikely to have the luxury of another quarter of losses—*your* boss will probably decide that you are not up to handling the issue and replace you.

Now, this is not to say that there are "no holds barred" where profits are concerned. Running a business is just like living your life—you must respect the law and deal ethically with your vendors, customers, employees, and stockholders. Managers who break the law usually wind up in court, and managers who treat others unethically quickly get a reputation that harms their business and their own careers. It may take a while, but people who are always testing the limits will find them, but usually not until it's too late.

Being really profitable—in the top 10% of your field—requires that you establish a leadership position in your line of business. A leader provides products or services that customers recognize as being superior to those offered by others in terms of *value*, and being willing to pay for them as such. This does not necessarily mean that your company must be the biggest in the industry or even in each product line. Sometimes being second or even third will allow you to concentrate on some specific area, such as a particular end use or a class of customer, where you can fully differentiate your offerings. Being a leader means that you always have new, high-profit potential products moving through the pipeline that will supplant the old ones when they become lower-profit commodities. It means constantly looking for ways to increase the value of your company to its customers, in terms of service as well as products. It means finding ways to differentiate your company from your competition, both direct and indirect. Only when your company is a leader can it increase profitability on an ongoing basis.

Around the turn of the past century, a number of financial scandals surfaced when several well-known, publicly held "Internet" companies declared bankruptcy after years of reporting constantly rising earnings. How could this happen? It turns out that questionable (if not fraudulent) accounting practices were used to hide losses or to report loans as sales revenues. We are not likely to see such manipulation go unnoticed in our industry because manufacturing is well known to be a cyclical business. Stock analysts would become suspicious if any publicly held plastics companies showed constantly increasing revenues and earnings, quarter after quarter and year after year, regardless of the business cycle. An honest and forthright presentation of financial and operating facts, blemishes and all, will create respect and credibility among investment analysts and the investing public.

2

Foundations of the Industry's Segments

There are key factors in each major plastics industry segment that must be recognized as strongly influencing the future success of every business. While some segments will have factors that coincide with others, each segment has at least one factor that is different from the others. These factors must always be at the base of strategic planning and business operations, or the company's building blocks will be weakened. This is not a matter of customer focus; it is knowing what elements are necessary to keep the business viable in the process of satisfying customer needs.

2.1 Polymer Manufacturing

Polymer manufacturing is a sector of the chemical and petrochemical industries as a consequence of historical development as well as the need for vertical integration. Virtually every polymer producer today is a division of a larger chemical or petrochemical company. Polymer manufacturing therefore must be analyzed from the standpoint of these larger industries. The production of polymers is a chemical process, unlike most downstream processing steps, which are physical processes. Polymer manufacturing is the most capital-intensive segment of the plastics industry, because the producer usually makes the monomers as well as polymerizes, so that the minimum plant scale to make a polymer is typically several orders of magnitude larger than for downstream processing of the same polymer. All of these conditions must be considered in order to understand how to manage a polymer manufacturing business. Most polymers are produced on a continuous, rather than a batch, basis. While this improves plant utilization and reduces capital requirements per unit of output, it also limits flexibility and requires more sophisticated operating controls, compared to batch processing. Batch processing, on the other hand, is more practical and less wasteful for making the smaller quantities typical of specialty polymers.

2.1.1 Technology

The most important factor to being successful in polymer manufacturing is (obviously!) *technology*. Without state-of-the-art technology to produce high-performance materials with consistent properties at competitive costs, no polymer manufacturer can remain in business for long. Let's review the three critical requirements of polymer technology:

- High performance
- Consistent properties
- Competitive costs

Each of these legs of the polymer manufacturer's stool is critical; without each of them, the stool collapses. Management must ensure that the company's technology is at least competitive or better, either through internal research and development, or through licensing, or a combination of both. Technology advancements include both product and process development. New products are vital to the growth of a company, but process improvement can make any product more profitable or enable it to compete against lower-cost materials in new end uses or both. Process improvement is also needed to meet environmental mandates to reduce waste generation, as well as reduce costs. Process improvement may even offer ways to modify the qualities of old products sufficiently as to make them significantly different than ones made with the standard process technology. Single-site (e.g., metallocene) catalysts are an excellent illustration of this point.

Patents are a vital part of polymer manufacturing technology. Not only do they provide protection of costly research and development, but they may also offer a source of income from licensing, as well as a *quid pro quo* to obtain access to others' patented technology via cross-licensing. Unfortunately, owners of valuable patents must expect that at some point it is likely they will be involved in litigation to protect their intellectual property. This is more common in the United States than in other countries, owing to the differences in legal systems. The first polypropylene patents were litigated for a period of more than 20 years before a final resolution was forthcoming. The winning party, Phillips Petroleum, won many millions of dollars in royalties as a result. The subject of patents and other intellectual property is examined in more detail in the next chapter.

2.1.2 Scale and Integration

One of the items mentioned earlier as a critical component of technology was competitive costs. However, technology is not the only way to achieve competitive costs; plant scale and integration also directly affect expenses incurred in manufacturing. Commodity polymers in particular must be produced in sufficiently large plants so as to minimize overhead costs. Integrated, on-site monomer-to-polymer production is another way to minimize cost by eliminating the expense of carrying duplicative inventories and transporting materials between sites. Integrated production also helps to ensure control over critical quality and cost issues, as well as to avoid supply interruptions. Logistics costs and currency exchange considerations also make it desirable to build integrated plants in various regions of the world in order to be able to supply customers profitably wherever they are located.

There are risks to integrated production, however. One is the risk that production outage in one stage will shut down all stages. The way around this is to have backup inventories at critical stages of production. A second risk is that the company must maintain its technology edge in *every* stage in the process, not just the one that brought it into the business in the first place, for example, polymerization. This means a bigger R&D effort, but it can also pay off by bringing in process cost improvements as well as being a defensive shield against competition.

A third risk is the bigger investment required for an integrated plant versus the smaller investment in a nonintegrated facility. The decision to put more investment at risk must be at least partially justifiable by relatively low probabilities of major changes in technology or shifts in the marketplace. Furthermore, as we have seen in the past half decade, increasingly stringent government regulation of plant emissions is becoming a serious cost concern. Since 2007, the US Environmental Protection Agency has undertaken a major push to measure and reduce *all* manufacturing by-products, whether solid, liquid, or gaseous (particularly carbon dioxide)—but simply implementing ever-more sensitive measurement protocols and procedures steadily drives up costs. If the facilities cannot meet the new regulatory requirements, one must decide which is the most cost-effective decision: to upgrade an otherwise serviceable but obsolescent plant or to tear it down and construct a more efficient, integrated facility, very possibly offshore, closer to markets with high growth potential. The Persian Gulf is a site where a number of very large, integrated polyolefin plants are located, with more under construction.

The ideal combination of size and technology has evidently not yet been found for polypropylene—Volker Trautz, former Chairman and CEO of Basell, told the author that the boom-bust business cycles over the years in this polymer have "effectively destroyed all investments made in the product since the beginning (the late 1950s)." This would seem to be more a problem of persistent overcapacity than finding an ideal plant size, however. Industry-wide overcapacity is ever a risk in commodity polymers, particularly as globalization has sharpened competition worldwide.

2.1.3 Routes to Market

Traditionally, polymer producers have utilized direct sales, distributors, and brokers as routes to the marketplace. Producers must use all of these routes to satisfy a diversified customer base. This is normally true wherever customers are located.

2.1.3.1 Direct Sales

Direct sales have the advantage of building and maintaining a manufacturer-to-customer relationship that has minimal "noise" or signal loss in communications. Direct sales representatives can be used to build large accounts over a period of time, something that is more difficult when using other routes to market. Direct sales relationships with customers are the strongest and the most reliable because the manufacturer is in direct control of the relationship without a third-party filter. Also, more individuals become involved through a direct sales business relationship than is the case with, say, a distributor; this helps to ensure that normal personnel turnover will not suddenly end contacts on each side. Additionally, changes in the marketplace are detected more rapidly and dealt with more effectively when working directly with customers.

Many polymer producers today use experienced technical personnel as "account managers" (the term *sales representative* is usually applied to junior personnel), who are able to deal effectively with problems on their own as well as knowing where and whom to go to in their own company for assistance. In line with the drive to minimize costs, such relatively expensive personnel are used to call only on the company's largest customers. Smaller customers are handed off to distributors. The dividing line for who will handle a customer varies with the degree of commoditization of the material. For example, a polypropylene processor who buys 3 K MT/yr. (thousand metric tons) might be a "house account," with smaller users being referred to a distributor, while a nylon processor who buys 450 MT/yr. may qualify as a house account. Other attributes that affect the decision on whether or not to designate a customer as a house account could include a new line of business, or sales potential where the company has only achieved second-supplier status. Also, customers who seek to develop new applications—a number of which will often require product development on your part—are served best through a direct selling relationship.

2.1.3.2 Distributors and Brokers

Distributors and brokers have the distinct advantage of not being a fixed cost. They also present a convenient and rapid way to move inventory off the books and into the marketplace, to be sold to the myriad of small processors that are very difficult for a manufacturer's direct sales force to handle on a cost-effective basis. Most polymer manufacturers have turned over all less-than-truckload or even less-than-carload buying accounts to distributors. Producers do this not only because it is more economical but also because they can instantly access a much larger number of potential customers through this channel than directly. The disadvantages of using distributors include that (a) knowledge of the marketplace is more difficult to obtain and less complete by going through these third parties, (b) gross margins may be adversely affected by turning over too much volume at reduced prices, and

(c) application development is typically not a strength of distributors and you risk missing new application development opportunities at customers served by distributors.

2.2 Compounding—Key Factors

Production of polymer is only one step toward creating a useful finished part. Often, polymers require modification of their properties by incorporating reinforcements, fillers, colorants, and additives to be successfully used in more demanding or unique applications. This step, compounding, is usually performed by companies other than the ones that made the polymers. Compounding, as a business, was first developed in North America by entrepreneurs; later, polymer producers began compounding their own resins, too. In Europe and Asia, compounding has been an integral part of polymer producers' business almost from the beginning, but independent companies have followed later. One can even observe small "part-time" compounders in some Asian and European countries, running only on a periodic basis to satisfy a few local customers. While this phenomenon is not a widespread pattern, it does seem to represent the ultimate in providing customers with just-in-time shipments of specific compounds at a lower cost than would be usually charged by a polymer producer or a full-time compounder.

2.2.1 Technology

As in polymer manufacturing, technology is a key factor to success in compounding. In particular, formulation technology is critical to finding and holding customers. Compounders must be willing to develop and manufacture special grades in the smaller volumes that polymer producers cannot or will not undertake. Customers for special grades usually want them for one of two reasons (often both!): (1) replace a more expensive material, for example, a flame retardant polypropylene to replace an engineering plastic, or a recycled product to replace a prime one; (2) provide a particular, even unique, combination of properties, for example, low friction and electromagnetic shielding.

Process technology is also important, and this usually involves the use of twin screw extruders in addition to single screw extruders. While it is hard to beat the economics of single screw compounding, there are a number of compounds that require the additional dispersion that is best achieved in a twin screw extruder.

Much of compounding technology, particularly processing, is in the public domain. In other words, it is widely known in the industry and the subject of published articles, so that it cannot be considered patentable or in the nature of a trade secret. However, unlikely or unusual *combinations* of public domain technologies *can* be deemed to be trade secrets. If this is the case, then a company can designate those portions of its technology that qualify as trade secrets and take the necessary precautions to treat it as such. This includes restricting access to the technology to a limited number of employees (no outsiders allowed!), having those employees sign secrecy agreements, and reminding those employees at least annually that the technology is a trade secret and the intellectual property of the company.

Many compounders use the trade secret route to protect themselves against employees leaving and using what they have learned by either joining a competitor or setting up their own competing company. A few companies (see Chapter 10) have patented their technology. While patents have the advantage of being simpler to enforce in court than trade secrets against competitors, they do have the disadvantage of transferring everything disclosed into the public domain after their term has expired. Therefore, embarking on a patent program requires an ongoing effort to continuously develop and patent *improvements* on the basic technology, so as to maintain ongoing protection.

2.2.2 Supplier Relationships

While supplier relationships may not seem to be critical to some small compounders, they are actually very important to achieving significant, sustained, and profitable growth. The largest single cost of doing business as a compounder is raw materials. Therefore, a close working relationship with one or more polymer manufacturers is key to securing a long-term supply of consistent-quality materials at an attractive price. Think of it this way: for the most part, the market determines the price at which one can sell products; therefore, a major factor in profitability is how well costs are controlled, particularly the largest one: materials. Polycarbonate is a good example of the need for a long-term working relationship. It has been in tight supply a number of times in the past three decades—with a simultaneous run-up in prices—suggesting that having more than one secure source of polycarbonate, preferably with contractual pricing and quantities specified, is critical for any compounder with a steady business based on this polymer.

2.2.3 Geographic Dispersion for Customer Focus

While most compounders start out with one location that serves customers in the contiguous geographic region, eventually they find that growth will require another manufacturing facility, located closer to distant customers. This is because just-in-time (no stock) customers can't wait for shipments overnight and sometimes not even for more than half a day. The very largest compounders also find they must follow their customers all over the world and build regional plants to serve these needs. The saying about politics—"it's all local"—applies equally to serving customers. While some major customers may buy globally, all of their plant sites must be served locally.

Once overseas plants are established, some technical capability must also follow, to qualify raw material sources, handle technical service needs, and handle simpler new product development requests. This is also an important way to gain new business in the region, in addition to serving the local plants of established global customers.

2.3 Distribution—Key Factors

Distribution is an important route to market—almost all small processors buy their materials from distributors because of the substantial minimum order size requirements of polymer producers. Distributors smooth out disruptions of the supply chain by stocking various grades of materials, enabling small processors to minimize their own inventories and polymer producers to maintain long production runs. Brokers are a small subset of the distribution business: they usually buy and resell surplus stocks rather than representing one or more specific manufacturers and stocking their products. Brokers sometimes do not even take title to the goods sold, receiving a commission from the party requesting the transaction.

2.3.1 Customer Relationships

Most distributors start out in business with several key customers—people that the founder knows well enough to be assured that they will buy products from the founder, in such quantities, prices, and degree of regularity as to ensure a profitable business. A sound customer base is an absolutely fundamental requirement for any distributor.

As noted below, a relationship with polymer manufacturers and/or compounders will also add to the customer base. There is a difference here, though. These customers are loyal primarily to the manufacturer/compounder, rather than the distributor. Should the supplier relationship end, these customers will usually switch to another source of the supplier's products. Therefore, the wise distributor will try to build a business relationship with these customers that are "on loan" from the supplier, so that they remain loyal for at least some of their purchases even if the supplier decides that the distributor does not fit in the business model any longer.

The old "80/20 rule" seems to apply overall to plastics industry buyers; that is, 80% of product sales are purchased by 20% of the customers and vice versa. The rule is modified a bit when it comes to the human need for trust when dealing with another versus the need for finding a bargain. Here we find that perhaps 70% of purchases are made predominately on the basis of personal relationships and 30% are made predominately on the basis of price. That 70% is where a distributor's important earnings are made. A distributor earns a customer's trust by delivering the right material on time every time, delivering paperwork, for example, invoices, material safety data sheets, certificates of analysis, which is accurate and timely every time, reacting to inquiries and problems immediately and helpfully. Much more often than not, this kind of reliability has enough "added value" to a customer to be worth paying a price premium. Old-fashioned "schmoozing" (entertaining the customer) may help cement a personal relationship, but without the foundation of reliability, it can be overturned by a competitor very easily.

Distributors need to know what else their customers buy and, if feasible, find a way to offer these additional products, too. This does not mean to mindlessly expand the product line, but to look for opportunities to carry related products that can be profitably sold to existing as well as new customers. For example, if the customer is already buying brand X nylon 6 but also uses brand Y polycarbonate, the distributor should try to establish a supply position in brand Y. The cost of selling and delivering two products to a customer is virtually the same as selling one product, so the additional profitability potential is obvious.

Restructuring at polymer producers has lead to cutbacks in technical service for customers. Distributors can see this as an opportunity to cement customer relationships by providing this support, usually in the form of helping solve processing problems. Technical service personnel do not have to be salaried employees, sitting around waiting for calls for help! There are many good consultants in this field, who can be brought in on a case-by-case basis as needed. If a consultant is paid a retainer, he or she is often willing to make him or herself available on short notice to handle emergency problems by telephone or even in person. Some common but less urgent technical problems can be handled via an automated telephone troubleshooting system, a fax-back system, or an interactive website.

2.3.2 Supplier Relationships

A distributor without one or more regular supplier relationships is more properly described as a broker. Supplier relationships are critical to continuity and security of supply, customer referrals, and operating profit margins. A distributor may have started out with a number of customers, but to grow on a sustained basis, a formal relationship with one or more suppliers is essential. In the ideal supplier relationship, the distributor is treated as a marketing and logistics extension of the polymer producer or compounder. Most polymer producers will refer all less-than-truckload customers to their distributors. Suppliers provide product literature and training to distributor technical personnel so that, in turn, the distributor can provide product information and technical service to small customers. Most importantly, suppliers provide a discount on purchases that can constitute most, if not all, of the gross profit for a distributor.

The ideal supplier relationship is to be the exclusive distributor in a given market, be it geographic, market, or any other basis (the broader, the better, of course). The next best is to be one of a very few. For example, during the 1990s, Shell licensed only three North American compounders to make and distribute its Kraton™ thermoplastic elastomer (TPE) compounds, which was very nearly as good as being an exclusive arrangement. A written agreement is to be greatly preferred over an oral agreement, and is essential when the relationship has restrictions. At the bottom of the list are nonexclusive distribution agreements, but they are often necessary to round out a product distribution line; it is not uncommon to make oral nonexclusive agreements.

Commodity polymer distributors will typically repackage bulk resins for customers who do not need to buy full railcar quantities. The distributor receives railcar shipments from the polymer producer and either puts the polymer into bulk trucks or into silos, where the polymer can be repackaged in bags or boxes. This is an important function in the supply chain for small- and medium-size processors. The distributor not only provides the desired packaging, but also ensures continuity of supply at stable prices. Some commodity polymer distributors delegate less-than-truckload business to sub-distributors of their own.

2.3.3 Geographic Dispersion

Distributors typically grow from local businesses to regional ones. A few have grown to be national or even international in coverage. It is difficult to grow solely in a local area for long, and therefore geographic diversification of customers becomes important to the continued growth of the business.

Another significant reason for geographic diversification is that some local areas tend to be dependent for their commercial well-being on specific large industries or companies. For example, southeastern Michigan ("Detroit") is heavily dependent on the automotive industry, Seattle/Puget Sound on aerospace (Boeing), and San Jose ("Silicon Valley") on electronics. It's a rare industry that doesn't have some down moments over the years. If the distributor's home market is one of these cities, then it should be doing its best to diversify its business base by developing customers in other localities.

Suppliers—and customers—are more favorably disposed to work with distributors that have broad geographic representation than those that are only active in a few areas, and usually is a requirement for processors with multiple plant locations. This should never be a problem for a distributor, as it is not necessary to own a number of warehouses. There are many perfectly adequate public warehouses that can be very effectively utilized on an as-needed basis, without the need of owning, staffing, and managing substantial real estate assets. For example, the Denver area is home to a number of mostly small processors, whose need for overnight deliveries can be best served by keeping stocks in local public warehouses.

2.4 Processing—Key Factors

Processors cover a wide range—from tiny, "garage shops" to billion-dollar multinational companies, from custom parts producers to original-equipment manufacturers' captive operations. They include specialists in different types of processing, in particular, classes of materials, specific types of parts or industries, etc. Some just pass plastic through a machine, some also make molds, others also decorate parts and assemble them. Some have become complete "contract manufacturers," with an emphasis on plastics. However, there are key factors that apply to each of these diversified entities.

2.4.1 Technology

The foundation of any processor is—again—technology. A selected process is used to transform polymers or compounds into functional parts or even completed objects. Processes include injection molding, extrusion, blow molding, rotomolding, thermoforming, compression molding, and some variations or combinations of the preceding types. Processors must have more than basic competence in the technology of the processes used or they will not stay in business for very long.

Many processors do more than make parts out of polymers and compounds. Their technology base often also extends to product design, mold or die design and construction, and secondary processing, for example, assembly, decorating, etc. These additional capabilities are important elements of broadening their customer base and improving profitability. These competencies should never be simply "me too" efforts, but be every bit as cutting-edge and high-quality as the processor's equipment allows. Even such a simple item as the right type of resin dryer is critical to making quality parts from hygroscopic polymers.

2.4.2 Customer Relationships

Processors may have one customer, such as a captive operation, or many, as custom processors do. Processors' geographic proximity to their customers is often the single most critical consideration for customers who practice *kanban*, the Japanese word for "just-in-time" inventory management. Just as in polymer manufacturing, compounding, and distribution, processors must diversify their customer base, to ensure that their future is not irrevocably tied to the fate of a single customer, however great that customer is. Some captive operations have entered the custom processing business, to utilize otherwise idle machine time and improve their internal profitability.

Processors are also finding that they need to offer more than just machine time in order to keep their customers happy. Adding additional services is not just a route to increased profitability, it can also mean survival. Some processors have gone so far as to characterize themselves as "contract manu-facturers," offering design, production, stocking, and shipping.

2.5 Equipment, Additives, and Others

Firms that offer equipment, additives, and the like to the industry are important contributors, but there are a number of relatively small specialized companies, as well as many small specialized divisions of much larger companies, whose primary business may well be in other, non-plastic markets. As warned in the preface, the variety of products, companies, and interests under this heading preclude anything more than limited commentary.

2.5.1 Technology

It may seem repetitious, but it is technology once more that is the reason that these firms exist. The need to process materials more efficiently, to endow compounds with enhanced properties, to measure those properties accurately and reproducibly, etc., is what drives the business of this group of companies. The plastics industry could not exist in its present form without these vital technologies.

2.5.2 Critical Mass

One observes that all of these segments of the industry are evidently subject to much the same cost pressures. Smaller companies are acquired by larger ones in order to achieve the minimum size necessary to compete on a broad scale, both from the standpoint of product line as well as global presence. The use of the combined products in diversified markets helps to spread overhead over a larger sales base and mitigates economic cycles to some extent.

Consolidation has particularly affected the machinery makers since the turn of the century, due to the two economic recessions in this period (2001–2012); the number of molders in the United States, for example, has fallen sharply by over one-third, flooding the market with used machinery and driving new machinery demand sharply down. Consequently, there have been a number of mergers that have rolled up many older European and American brand names into significantly fewer companies. The latter will be better able to withstand the challenge from Chinese machinery companies, which are now emerging as formidable competitors.

2.5.3 Customer Relationships

Yes, good relationships with customers are important for these industries, too, especially equipment suppliers. When manufacturers expand or upgrade, the performance of the equipment previously purchased and the service received will mean at least as much, if not more than, the price paid, when it comes to the next purchase. This is frequently a matter of *when*, not *if*, so that the supplier must keep in touch with the customer even though there has been only a single purchase made and that more than a year ago. Despite their specialized nature, these suppliers exist in highly competitive markets. Companies in this industry segment find that most of their products are sold in limited quantities and at relatively lengthy intervals. This means that direct sales must concentrate on the largest customers and using distributors to handle smaller ones.

3

Technologies and Markets Shape a Company's Business

As mentioned earlier, the plastics industry is dividing into product and market segments that strongly affect how companies have to run their businesses. In a number of cases, the company's original management did not consciously decide to be in certain lines of business, but the company's technology and markets have led them there. Technology is the one common key factor that connects almost every one of the different industry segments. You need to be aware of these powerful influences and take them into account when managing the business.

3.1 Technologies

3.1.1 Materials

The technical characteristics of materials produced, compounded, or distributed by a company, characterize and drive the form in which its business is conducted. A commodity producer will have great difficulty in successfully operating its business as though it made specialty materials, and vice versa. Managers ignore this principle at their own risk—and a surprising number do, and with unsurprising, sorry consequences! Figure 3.1 shows an estimate of the physical volume in pounds and value in US dollars of individual commodity polymers, a semi-commodity (ABS), and engineering polymers (as a group), sold in the United States and Canada in 2011.

3.1.1.1 Commodity and Semi-Commodity Materials

There is a simple correlation between market price and sales volume, as one might expect. Higher prices correspond to less volume and vice versa. Of course, this is after the product has reached some degree of maturity—market forces take a while to react to price changes. Lower-priced materials are, for the purposes of this discussion, classed as commodities. Commodity resins are usually defined as including polypropylene (PP), polyethylene (PE), polystyrene (PS), polyvinyl chloride (PVC), and polyethylene terephthlate (PET). A number of materials usually thought of as engineering plastics have taken

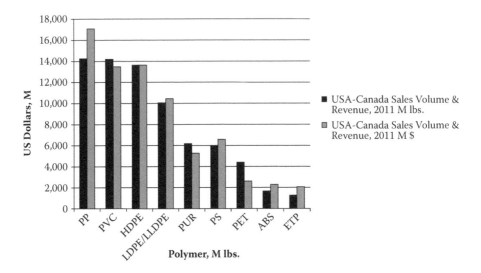

FIGURE 3.1
USA–Canada estimated sales volume and revenue, 2011.

on semi-commodity characteristics, such as significant price fluctuation. Among these are acrylics (PMMA, both thermoset and thermoplastic), acrylonitrile-butadiene-styrene (ABS), styrene acrylonitrile (SAN), nylons (PA), polycarbonate (PC), and acetal (POM). A number of thermoset materials may also be classified as semi-commodities, such as phenolics, urethanes, alkyds, unsaturated polyesters, aminos, and epoxies.

What are the principal characteristics of commodities and semi-commodity materials? They are, first, largely interchangeable within families and grades; i.e., virtually every supplier's 12 melt-flow homopolymer polypropylene resin can be used in a molding machine with few, if any, apparent differences between these products with respect to processing and physical properties. Second, suppliers usually compete with each other primarily on price, rather than on property differences. These factors are seldom true with respect to many engineering polymers; one manufacturer's product does not necessarily match up across the board with another's seemingly equivalent product. Third, the physical volumes of commodities to be manufactured, transported, stored, and processed are entire orders of magnitude larger than those of semi-commodities.

What are the requirements for managing a business based on such materials? Management has to focus on cost control while finding ways to develop differentiated products. Cost control in the case of commodities usually means minimizing the number of grades produced, maximizing the length of production runs and size of orders—the reverse of specialized products. Commodity materials are, by definition, volume materials. Logistics play a critical part of cost as well as customer service. Bulk shipping is the usual rule for commodities, which requires an investment in transport, storage, and strategic plant siting. These considerations are, at best, only occasionally found in semi-commodity materials.

Even sales personnel are affected by these factors. Commodity sales representatives have to concentrate on the largest users and try to obtain long-term contracts. Semi-commodity sales representatives are inclined to spend more time looking for new applications or qualifying as a second source for an established and growing application, less emphasis on sales volume, and more on unit and account profitability.

3.1.1.2 High-Performance and Unique Materials

High-performance materials are usually the most expensive of all plastic materials and feature one or more outstanding properties for a polymer material, such as exceptionally high resistance to heat and chemical attack, high dimensional stability over a wide temperature range, or high lubricity. This group usually includes polysulfones (PSUs, PESs, and PPSUs), polyetherketones (PEKs, PAEKs, PEEKs, etc.), polyphenylene sulfide (PPS), liquid crystal polymers (LCPs), polyimides (PAIs, PEIs, etc.), thermoset alkyls, and silicones. Unlike commodity materials, high-performance products that are nominally the same may differ markedly from each other as to properties, processing characteristics, and price. For example, LCPs from DuPont and Ticona vary significantly from each other in the characteristics just mentioned. Because these materials are relatively expensive, they are sold in much smaller quantities than commodity materials. Therefore, logistics and customer service, while very important, are usually less significant factors in terms of cost per unit sold. High-performance materials are packaged in bags, boxes, or sacks, and virtually never sold in bulk. Creating a business in these products requires a strong product and application development group to keep sales and earnings curves moving up smartly.

Few materials are truly unique. All plastic materials compete with each other and conventional materials, at least to some degree; however, it is the combination of properties and cost that ultimately decides which material will be used in any given applications. Even those protected by patents must compete with others that overlap their properties in some way.

As an example of a unique material, consider polymethylpentene-1 (PMP). While its optical properties do overlap those of some other transparent polymers, its combination of gas permeation, light transmission, heat, and chemical resistance set it apart from other transparent polymers such as acrylic, PET, PC, and SAN. PMP's relatively high cost keeps it from taking away more than a small number of high-end applications from the competing products, but this pricing must be considered a deliberate choice by the manufacturer to maximize profitability in markets where sales volumes are limited.

Polytetrafluoroethylene (PTFE) is another relatively unique material, both from the standpoint of its properties and significant sales volume despite its relatively high selling price. Nevertheless, it has become a semi-commodity in the course of its nearly 80-year commercial history, considering the near-uniform range of grades offered by producers. PTFE has a remarkable combination of chemical inertness, lubricity, ignition resistance, and dielectric properties that are not found in almost any other polymer. However, it is not melt-processible and, consequently, PTFE fabricators constitute a distinct group among plastics processors, who use techniques similar to metal fabrication. Melt-processible fluoro-copolymers, such as fluorinated ethylene-propylene (FEP), and perfluoroalkoxy (PFA) are more expensive than PTFE but retain most of the beneficial properties of PTFE, as well as being melt-processible (but are significantly more expensive).

Pricing unique materials is a distinct challenge. One cannot ignore other materials that come close in properties, because too big a price differential can allow a competing product to gain a foothold at the low end of some applications, or encourage the user to redesign the part to use a less expensive material. Producers of unique materials have found that application development is essential to keep unique product sales from falling to GDP growth rates when saturation of their initial markets is reached.

3.1.1.3 Support Requirements

Each type of material requires different levels and types of support. For example, commodities require the least technical support, while technical and high-performance materials require the most. On the contrary, commodities require more investment in logistics than do high-performance materials. The impact on the business is subtle—increased R&D support is a direct expense, and normally runs between 2% and 5% of sales, whereas increased logistics support is likely to be in the form of additional investment in storage and transport, showing up in financial statements as depreciation and maintenance costs. Depreciation must be written off over a period of years, which tends to conceal its real cost (and increases taxable income). While some companies do capitalize and depreciate R&D expenses, this is usually avoided unless the work results in a patent.

3.1.2 Processing Equipment

Processors' businesses are also affected by technology; in this case, it is the equipment they use. Not only equipment types, but also the scale and range of integration come into play when considering the technology impact on the nature of processors' business models.

To a large extent, a processor's equipment defines the business of that processor. A molder with 2500 ton clamp presses is more likely to be involved in business machine or automotive markets than in power tool

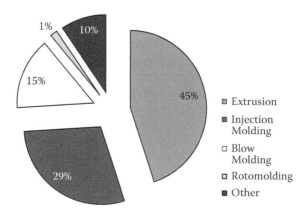

FIGURE 3.2
Estimated polymer consumption by process, 2011.

markets, just as a molder with 75 ton clamp presses is more likely to be supplying electrical/electronics than large appliance markets. A rotomolder is more likely to be involved in consumer products than in medical equipment markets. And a pipe extruder is almost by definition supplying the construction market.

As mentioned earlier, some processors specialize in types of materials processed, such as thermosets, or polytetrafluoroethylene (PTFE). Again, this tends to define the markets served, as well. Thermoset molders are likely to be focused on electrical/electronics and PTFE processors on chemical process equipment and some automotive components.

Figure 3.2 shows an estimate of the relative volumes of plastics materials processed in the United States and Canada in 2011, by type of equipment.

3.1.2.1 Equipment Types—Opportunities or Limitations?

The answer to this question, of course, is both. Injection molders seldom compete with extruders, although blow molders may find they are competing with thermoformers or rotomolders. Equipment type and size often give processors a chance to be a specialist or a generalist, as well as a combination of both. A range of machine sizes is more likely to be found in the plant of a large general custom molder, for example, than in the shop of a small custom molder. In the global marketplace, owning large equipment and making large parts can contribute to a more secure business base; the speed and lower cost of delivery to customers is likely to offset enough of the low-price bids by offshore molders to secure business; local inventories are usually smaller than those required when buying offshore, which is another plus.

3.1.2.2 *Full Service versus Specialist*

As processors grow in size, they will usually come to a point where they find that their initial focus on a single process limits future growth and/or profitability. Often the decision to diversify equipment and processing capabilities is forced by a major customer who wants "one-stop shopping." The processor ignores such desires at the risk of losing most or even all of that customer's business. On the other hand, not every added service adds value; don't make the mistake of pricing some services below cost on the theory that you will earn more from additional overall business. That is not sound practice. For example, if a customer wants the molder to electroplate parts, this should be analyzed as a "make or buy" situation—whichever course makes a more satisfactory return on investment will be the one to take—it should not be an automatic reaction to go out and buy electroplating equipment. Buying equipment that will only be used to support one customer presents a risk that should be appraised carefully. If you can pay off the cost and make a profit within the life of the purchase contract offered by the customer, then this may well make sense. But if the process is unfamiliar to you or the equipment unlikely to be used more than a fraction of the time, it may well be more beneficial to contract out the work involved.

Nevertheless, some forms of integration can reduce processing costs and are properly viewed as *productivity* enhancements, rather than business diversification. An example would be Pushtrusion® and similar processes that make long-fiber reinforced thermoplastic parts from continuous fiber roving and polymer pellets, rather than buying LFTP pre-compounded pellets for conventional injection molding. The added equipment cost mandates using such an approach for long runs of large parts, so not every molder will find this type of process to be useful or economical. So far, the principal use of this technology has been for producing large automotive parts.

3.1.3 Patents, Trade Secrets, and Licensing

As noted in Chapter 2, technology is generally a critical factor for competing successfully in the plastics industry. Technology comes in several varieties:

- *In the public domain*, meaning that the technology has been published in the open literature and is free of any patent restrictions. Most basic manufacturing operations fall in this category, such as simple molding, compounding, or polymerization steps that follow long-established processing technologies and/or well-known chemical engineering "unit operations." Most basic plastics processing technologies are in the public domain.

- *Patented*, meaning that the composition of matter or process or usage has been described in detail in a patent that has a finite life, after which the technology enters the public domain. Only the owner of the patent may use the technology or grant others the right to use it. Anyone infringing a patent (using the technology it claims) without permission may be sued by the patent owner for damages, as well as a "cease and desist" injunction requested from the court for immediate relief. Patent lawsuits are among the most expensive of civil cases to prosecute or defend and the financial stakes must be high enough to justify the costs. Such legal actions also require a substantial amount of time from company personnel for the preparation and conduct of the lawsuit, as well as dealing with possible appeals. Companies do well to be sure they are not using any patented technologies without written permission from the owners.

- *Trade secrets*, meaning technology that is both unpublished and not generally known or practiced outside the company that uses it. The company that claims ownership of a trade secret must take steps to ensure that it remains proprietary, such as restricting access to the formulation or process area where the secret is used, and requiring every employee to sign a secrecy agreement that he or she will not disclose or use the secret outside their employment with the company.

Trade secrets and patents may be licensed to others, and it often makes sense to do so. For the licensee, this route offers fast access to proven technology without the cost, risk, or delay of having to develop comparable technology. For the licensor, this route offers an additional source of financial returns on the investment it made to develop the technology, and, often, access to any improvements made by the licensee. While it is true that by issuing one or more licenses, the licensor may increase competition for itself, this can actually be advantageous in some situations. The reason is that many potential large customers for the patented or trade-secret-protected product may choose not to use a single-source material, out of concern for sufficient supplies or the monopoly power of the manufacturer to set artificially high prices or both. Introducing a second supplier, even though under license to the first supplier, usually removes these concerns and causes the demand for the product to grow much more rapidly than would be the case with only one supplier. Also, second suppliers frequently develop new applications and markets faster than one, not only because they bring additional assets to bear, but also because it makes more business sense to find opportunities where the initial supplier is not active.

Finally, patent holders have to recognize that their monopoly has a limited life and they can gain more during the lifetime of the patent by licensing to create a strong duopoly that will make it more challenging for others to enter the business after the patent has expired. Unfortunately, not many patent owners have been willing to observe and learn from the few who have pursued limited licensing for additional income or used it as a way to end expensive patent infringement litigation. DuPont and Celanese chose this course after briefly contesting each other's POM patents (they cross-licensed each other), and both found that their POM businesses prospered afterwards.

Often, engineering firms or equipment suppliers will furnish technology license packages as part of their products and services. "Turnkey" plant components and layouts are usually based on information in the public domain. Individual equipment items may be patented or treated as trade secrets, but the buyer gets a license as part of the purchase.

3.1.4 Regulatory and Environmental Issues

Most polymers, particularly such commodity materials as polyolefins and polyethylene terephthalate, are chemically inert and therefore relatively benign, from an environmental standpoint. However, PP, PE, and PET are so widely used for consumer goods packaging, that they pose a litter problem when thoughtless end users discard them. Management needs to actively promote recycling or incineration of these materials as much as possible. Nevertheless, the economic picture for recycling versus incineration must be addressed on a long-term basis via industry associations since local, state, and federal government mandates seldom consider life-cycle costs to be an issue for discussion.

Polyvinyl chloride (PVC) has received particular opprobrium from some environmental groups: the monomer, vinyl chloride, is a known human carcinogen, and partial incineration at low temperatures (a remote likelihood) of the polymer can lead to the formation of certain dioxins that have been shown to cause dermatitis in humans and cancer in guinea pigs. Despite this adverse publicity, it is clear that PVC can be successfully recycled or incinerated under normal conditions. Nevertheless, recycling consumer-source PVC seems unlikely, since collection and sorting by municipalities is largely limited to PET and PE, although interest in recycling PP appears to be increasing.

Engineering and high-performance plastics can be recycled, but economics favor postindustrial rather than postconsumer sources, in order to have an identifiable and relatively clean waste stream of sufficient size. For example, nylon fiber waste is an ideal source for recycled nylon molding and extrusion compounds. Most processors recover sprues, runners, and scrap parts as part of their normal operations; if recycled material is not permitted to be used in making parts, then the "regrind" material can be sold to brokers or other processors, where it ends up being recycled.

Another promising approach is to use heat and pressure to reduce waste polymers back to crude oil; several test programs are in progress to demonstrate commercial potential. In any event, it is wasteful and a poor business practice to send plastics scrap of any type to landfills.

Thermoset materials can also be recycled in the form of filler for use in virgin compounds. This has been successfully demonstrated in polyester and epoxy molding compounds; there would not seem to be any technical reason why other thermoset resins cannot be recycled in the same manner.

Many processors do not have the background in chemistry or chemical engineering to be knowledgeable about the dangers of toxic or dangerous fumes coming from overheated or decomposing polymers. This information is readily available from suppliers, often in simpler form that that in Material Safety Data Sheets (MSDS), which tend to be quite technical and legalistic. In addition to technical support from suppliers, processors will find that industry trade associations, such as The Society of the Plastics Industry, can be helpful in identifying such problems and the proper steps to deal with them effectively and safely.

3.2 Markets

The enormous variety of end uses for plastics materials is what gives the industry its truly dynamic character. It also offers an exciting and defining challenge to find the optimum mix of markets and customers to pursue that fits the technology your company uses, the products you are capable of making, the services you offer, and your financial goals. It is worth noting that a surprisingly high percentage of people who worked in the plastics industry but left it for one reason or another, often return because they missed the seemingly endless variety of new applications and business opportunities.

While some companies have concentrated their efforts almost entirely on larger end use markets, many find it safer to spread their business over a broader variety of applications. As noted in the previous chapter, the types of products a company makes strongly influence its marketing strategies and efforts. Figure 3.3 shows an estimate of the relative size of different major market segments by physical volume of all types of polymers consumed in 2011.

3.2.1 Packaging

Packaging constitutes the single largest end use for plastic materials, primarily the commodity resins: polypropylene (PP), polyethylene (PE), polystyrene (PS), polyvinyl chloride (PVC), and polyethylene terephthalate (PET). Nylons (PA) are an important exception, because film for food packaging is a major market for these polymers.

Films and containers for consumer goods and food constitute most of this market and, for the most part, are truly commodities, using commodity polymers. Postconsumer recycling is also an important consideration for PE- and PET-based packaging; some government agencies, as well as some

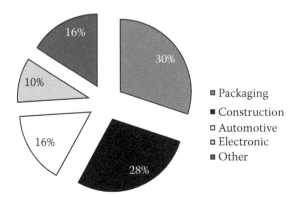

FIGURE 3.3
Estimated plastics market proportions, 2011.

major end users, specify that packaging resins have recycled content. One of the attractive aspects of the packaging market is its relative resistance to cyclicality, at least in food packaging. People have to eat, and food packaging is actually growing faster than the population increase, apparently due to the popularity of convenience foods.

There are also a number of specialty uses that are much smaller in volume, but offer better earnings potential to both the supplier and the user, such as industrial machinery and custom packaging. These uses can range from made-to-order polyurethane foam protective moldings to injection-molded acrylic cases for small tools. Despite its higher cost, PMP has found some niche food packaging applications, where its unusual combination of high transparency, gas permeability, heat, and chemical resistance, can offer greater value than the commodity polymers mentioned earlier.

The number of firms using polymers to produce packaging materials is substantial, but each company's usage is usually sufficiently large that they have significant purchasing leverage. Polymer shipments are frequently made via bulk carrier, either rail or truck, to these large users. Therefore, expertise in logistics is often critical to serving these customers. One should never forget, however, that "cost is king" with such users, and you will have to prove your ability to deliver value every day.

3.2.2 Construction

Construction is a distinctly cyclical industry, and most applications are very much commodity in nature. Much of the volume usage is processed by extrusion, for example, piping, conduit, siding, etc. This category can include agricultural end uses such as irrigation pipe and fittings. Building starts have been hurt badly by the 2007 recession and have been relatively slow to return to earlier levels. You may find that catering to the "do it yourself" market is more attractive in terms of demand stability than higher-volume general contracting business.

3.2.3 Automotive

The automotive industry is the single largest end user for many engineering plastics, such as nylons (PA), polycarbonate (PC), acetal (POM), modified polyphenylene ether (MPPE), etc. It is also a very important market for commodity polymers, for example, PP, PE, and PVC. The automotive industry is unlike the packaging industry in that there are relatively few users, and these try to limit the number of their suppliers, the better to cut costs by offering more sales revenue in return for lower prices. The sales revenue potential of the automotive industry is so large that many firms are attracted to it, but profitability is not only low but also under constant and heavy pressure from both customers and competitors. Automotive business is notoriously cyclical, and suppliers can easily find their orders cancelled literally overnight if there is a downturn in demand.

In the past, among US auto producers, General Motors has placed particularly onerous demands on its parts suppliers to reduce prices by a fixed amount per year, even retroactively, or face the risk of being phased out as a supplier. This has lead to a number of suppliers merging in order to reduce costs; several have entered bankruptcy. Other suppliers have either switched to supplying other automotive companies (typically non-US nameplate, such as VW or Toyota), or have sought to diversify their business into nonautomotive markets. General Motors acknowledged that these policies risked losing suppliers but was willing to gamble that others would take their place in hopes of winning a piece of the potentially large sales volumes at these firms. GM filed for reorganization under the bankruptcy laws at the beginning of the 2007 recession and was rescued by the US government; bankrupt suppliers were not so fortunate, and the industry remains in serious economic trouble. Nevertheless, suppliers to Japanese, Korean, and German automakers seem to be weathering the recession and weak recovery. These customers, although smaller in overall demand, appear to be more secure and rewarding to serve.

3.2.4 Electrical/Electronic

In the more distant past, this was everyone's favorite end use market. A large number of customers using a wide variety of materials (many custom-made) in small quantities for each application, yielded good revenues with excellent profitability and high growth rates. Globalization has changed this forever, as industry consolidation has reduced the number of end users, and outsourcing of production has lead to fewer actual manufacturers overall. These developments, in turn, have resulted in increasing competitive cost pressures keeping prices—and profits—down. In the past two decades, E/E buyers have overall been shifting manufacturing out of the United States to lower-cost countries, such as Mexico and Asia. Sometimes, polymer sales follow these shifts, but for processors, the best opportunities in this business area seem limited to start-ups and small specialty companies.

A unique aspect of the "E/E" market is the fast "time-to-market" demands of the industry—usually six months or less—and relatively short product life cycles, perhaps 12 to 18 months. These considerations virtually preclude multiple sourcing of materials and parts other than the most basic of units, such as connectors. They also mean that suppliers must work with end users with new applications from the beginning—for which they will be rewarded with the business—but they will be unable to displace or even share business with an existing supplier unless there has been a major quality failure.

Doing business in this market requires that R&D evaluate every new product development, both materials and parts, with both Underwriters Laboratory (UL) and European Union requirements in mind. This almost always means that materials must be offered in flame-resistant formulations if they are not already inherently flame resistant. Regulatory requirements demand that flame resistance be achieved without the use of halogenated components. Another important UL requirement is the temperature index, the maximum continuous use (operating) temperature. As electronic assemblies grow ever smaller, higher operating temperatures are the norm, with corresponding demands on higher-performance materials and part design.

3.2.5 Consumer Goods

Consumer goods can be divided into two categories: durables and nondurables. Consumer durables include household appliances, power tools, lawn and garden equipment, recreational goods, toys, etc. This market segment is widely diversified, with some very profitable niche opportunities. The downside of business in consumer durables is that the product life cycles are often short and many applications are highly price sensitive. While most of this market is national, the great majority of end users purchase not only from suppliers in the three North American Free Trade Association (NAFTA) countries—the United States, Canada, and Mexico—but internationally, particularly China. Many processors and material suppliers have followed these users overseas in order to serve their customer more effectively than is possible from a firm sited only in the United States.

Consumer nondurables consist mostly of packaging and single-use, disposable items. Again, cost is king, as one would expect in commodity applications. Since freight is an important component of delivered cost for packaging, there is less offshore competition for NAFTA-based business.

3.2.6 Industrial Components and Semifinished Shapes

Industrial components are a bit of a catch-all, but this end use is mainly concerned with machinery parts, such as pump housings and impellers, conveyor links, gears and bearings, etc. This is an ideal market into which to sell; it is highly fragmented, has low competitive visibility, high value-in-use and therefore relatively high potential profitability. Product life cycles are usually likely to be long. End users in this category are often local. Offsetting these advantages, volumes tend to be small.

Semifinished shapes are rod, tubing, and sheet, which will be mechanically fabricated into other parts for unidentified end uses. Often, semifinished shapes are used to make prototype parts for evaluation or even small numbers of commercial parts. This is an important market in terms of size, but its growth rate is less than many others, and it is very competitive. There has been considerable consolidation taking place in the last several years, so that the remaining processors are able to exercise more pressure on suppliers' prices than previously. This is an unusual business segment since shape producers rarely have direct contact with end users, their route to market being almost entirely through distributors.

3.2.7 Other

We needed a category for "everything else" and this is it! Nevertheless, there are a few identifiable components here, and some of the larger ones include the following.

3.2.7.1 Medical

The medical market is nearly as recession-proof as food packaging, but is growing faster. Medical products are of two types, disposables and durable equipment. Disposables are commodity products; the emphasis is on cost, but reliability and standards-compliance demands are much higher than for, say, houseware, making profit margins better than in many other commodity markets. Many disposable applications are relatively high-tech, such as catheters, blood bags, IV bags, etc.

Durable medical equipment offers the chance for higher margin business, since components are likely to require specialty materials and short molding or extrusion runs; reliability demands are also high, and you should review with your insurance carrier what your company's exposure to product liability claims may be and how to handle it.

3.2.7.2 Aerospace and Military

Aerospace and military markets are among the most challenging markets for plastics. Aerospace applications are generally decided upon when a new type of aircraft is designed, so there are relatively few companies and applications to pursue.

Military applications tend to be more dispersed than aerospace applications, however, often with relatively small potential volumes, plus a somewhat longer application development, approval, and procurement cycle. Fortunately, product life cycles tend to be relatively long. Both aerospace and military markets tend to be kinder to material suppliers than processors. Once a material specification is written around a particular product tightly enough, end users and your competitors alike will have difficulty justifying the time and cost required to qualify alternate sources.

Military parts contracts, on the other hand, are generally put up for bids periodically, so a processor has to defend the business regularly. Despite these limitations, a number of aerospace and military applications can offer a relatively steady business with somewhat less price pressure than is seen in many other end uses, because of the resistance to qualifying multiple sources. Consequently, the number of competitors tends to be limited.

4

Company Culture, Organization, and Direction

The plastics industry is still relatively young and growing, and it has a broad mixture of company cultures. As older companies of the industry grow in size or consolidate, or follow local customers to other parts of the country or of the world, their business cultures undergo changes. As a rule, polymer production has been parented by basic chemical or petrochemical companies; as a result, polymer manufacturing is usually associated with big company, managerial, and commodity cultures. Compounding, processing, and distribution, on the other hand, are most frequently found to have small-company, entrepreneurial cultures.

There is nothing intrinsically right or wrong about any particular type of company culture. They exist as a social phenomenon and must be viewed as such. However, one cannot manage a company successfully without a working understanding of its culture. The culture of a company defines how it reacts to stimulus, both internal and external. Sometimes this is appropriate to making a business grow profitably, sometimes not. Having assessed this, an executive may decide to either work within the existing culture or seek to change it. One way to change the culture is to revise the organizational structure and reassign some people to new responsibilities. You should understand that changing a culture is a slow, difficult process, with a significant potential for failure. However, once you are satisfied with the culture and organization that is in place, then you need to direct its functioning on an effective and efficient basis. This chapter deals with all of these considerations.

4.1 Size Matters—It's Intertwined with Culture

It's no surprise that communications, and consequently management, becomes more complex and difficult as the size of a company increases. Unfortunately, that's the good news. The bad news is that a number of people who function well in a small company often don't do as well in a large one and vice versa. Therefore, you must expect that there will be some personnel friction accompanying rapid growth or shrinkage, which goes beyond just staying even with the pace of business. The inflection points at which the size of the company affects its organization and the personnel in it will vary according to the function of the group involved, for example, manufacturing versus R&D. As a rule of thumb, new companies first feel these growing pains when they reach professional staffing levels that require two layers of supervision below top management. The first addition of a geographically separate site will also introduce some cultural friction, particularly if it is in another time zone, which tends to impede communications.

Company size influences company culture and vice versa. Small companies are more like small towns or even extended families, where working relationships are often also social ones. There are few secrets in small companies! Because of simpler organizational structure and fewer personnel, small companies are usually able to make fast decisions and react quickly to changes in the marketplace. Small companies generally have a strong customer focus. When they are not, it's often because they have a culture left over from a previously downsized, much larger company.

Large companies are typically more focused on global strategies and cost control, with particular emphasis on manufacturing and logistics. They are often, but not always, commodity oriented in some measure. As a rule, they are more overall end-use market or product focused than customer focused. Large companies also tend to be more impersonal, with more formal relationships between people working in different groups. Big company cultures can degenerate into bureaucratic, quasi-government cultures, where job security and turf protection can be a higher priority than company success. Frequent restructuring can produce this culture, as the employees begin to question management's loyalty to them and decide their best chance to survive such changes will be to assume a low profile ("don't make waves") and ride out the changes. Often, the customer's interests suffer when this situation is taking place.

In addition to being influenced by size, company cultures often reflect the dominant professional group, for example, technical or manufacturing or sales, when the company came into being or went through a reduction in size. Cultures also reflect the characteristics of the ownership, too, whether they are a founding family, institutional investors, or foreign nationals.

Why do we care about culture? Because it imposes certain constraints on how a company is managed. If management ignores culture and these constraints, it will have an uphill battle on its hands getting subordinates on board with its objectives and then reaching the desired performance goals. Changing a company culture is not a quick process without significant personnel resistance or turnover and the high risks that go with these reactions. The least risky—but the slowest—route to successful culture change is to accomplish it through one subordinate group at a time.

Some of the more frequently encountered cultural varieties are reviewed in the following paragraphs.

4.1.1 Entrepreneurial Culture

Entrepreneurial companies are almost invariably small ones, where the founder (usually the owner) is *the* boss, following his or her vision, and is typically involved in every detail of the business. If you are that boss, then you can make your entrepreneurial company an exciting and enjoyable place to work, if you structure the work environment so that everyone feels a sense of mission, participation, and accomplishment as the business grows and prospers. If you attempt to micromanage every detail, however, all you will ensure is that *you* will work 80+ hours per week and that your subordinates will grow frustrated as they are prevented from having a chance to do anything of their own undertaking. You will also ensure a high personnel turnover. In the normal business world, entrepreneurships generally either succeed or fail within two years of start-up.

Entrepreneurial companies are usually service and specialty products oriented, with the ability to tailor these offerings to the customers' needs. Virtually all of the entrepreneurial companies within the scope of this book will be compounders, distributors, and processors. The capital requirements for polymer manufacturing are generally so large as to preclude entrepreneurial startups. Early in the industry before scale of operations counted heavily, only a very few small companies grew big enough in molding or compounding that they could integrate backward into polymer manufacturing, for example, Foster Grant and LNP; these exceptions have never been repeated. Companies large enough to be polymer manufacturers quickly outgrow an entrepreneurial culture and transition to a managerial or commodity one.

Interpersonal relationships in entrepreneurial companies are usually very strong. This is unsurprising, since almost everyone has been hired by the founder and usually shares his or her sense of mission. The employees also usually exhibit a sense of loyalty to the founder and the company that gives them the incentive to work harder (if not smarter!).

Entrepreneurial companies are also often family-owned and managed. Although Andrew Carnegie thought that succeeding generations never measured up to the founder ("From shirtsleeves to shirtsleeves in three generations"), there are quite a few small- to medium-size companies that have flourished under family owner-managers, and even large companies, such as Huntsman Chemical (the Huntsman family). Perhaps the greatest strength found in such entrepreneurial companies is a continuing clear and consistent vision of the company's purpose in business, in contrast to many large, publicly held corporations.

The chemical industry in particular, in the past two decades, has seen some of its largest firms decide they would become life science companies, divesting some or all of their base businesses (such as plastics) and then realizing—too late—they were not big enough or did not have the requisite technology base to compete successfully against existing life sciences companies.

The result has been the disappearance from the plastics scene of some of the industry's oldest names, such as Hoechst and Monsanto.

Another benefit of family ownership can be the assurance that the firm's owners are committed to the company's growth and continuation, thereby giving some measure of security to the employees. Nevertheless, these decisions can be subject to change when generational succession takes place. Even Jon Huntsman, admitted that his family pushed him to take the company public, even though he had gone on record ten years previously that "... commodity chemicals [are] no place for the investing public. The cycles are too deep, the basic factors governing business are 75%–80% outside the control of managers, and investors don't understand that these types of businesses have periods in their cycles when business is so weak, dividends can't be paid. It's OK if a commodity business is part of a much larger company, because the impact is lessened. But it's far better that a commodity company be private. It can take the up cycles with the down." Huntsman, a cancer survivor, noted that taking Huntsman Chemical public was an important consideration in the course of his estate planning—a problem that confronts every successful entrepreneur who wishes avoid a forced sale or breakup of the company in order to pay "death taxes."

4.1.2 Managerial Culture

A managerial culture is frequently the successor to an entrepreneurial culture. By definition, the company's founder is no longer trying to run everything as he or she would in a start-up. The focus is on growth and profitability, rather than on becoming established and surviving. The management now views the company as part of their career, not as an extension of their personality. Or at least management *should* do so; sometimes senior managers can begin to think of the company as "theirs," and the company reverts to an entrepreneurial culture when it may not be appropriate to the situation. Interpersonal relationships and company loyalty are not as strong as in the entrepreneurial company. Some family owned businesses move into this category when the founder retires or dies, and the family wishes to retain ownership but to employ nonfamily management to run things. Managerial cultures not only exist in publicly owned companies, they are usually the prevailing one when a plastics industry company reaches $200 M or more in sales.

4.1.3 Commodity Culture

Commodity cultures are usually found in large companies (over $1 B in sales), with the focus on manufacturing and cost control cited earlier. For our purposes, commodities may be conveniently defined as those polymers that are produced in high volumes, such as polyolefins, styrenics, polyesters, and vinyl. Further, the properties of one commodity producer's primary product line are largely undifferentiated from other producers' equivalent products. This lack of significant differentiation between products normally means that competition keeps the prices of these materials down. Low prices mean that logistics and transaction costs are significant, and that operating profit margins are thin on an absolute basis. For these reasons, commodity companies emphasize long, smooth production runs, a minimum number of grades, and tight control over costs. When effectively managed, these companies can be very profitable. Over the decades, their biggest problem has been to avoid following the business cycle by overbuilding capacity just before a dip in the economy takes place.

Commodity culture companies tend to have flat, lean organizations. Interpersonal relationships and company loyalty are important to the successful operation of such businesses where so much depends on so few people. Unfortunately, some commodity companies may lack such loyalty if they have gone through a number of staff reductions in the course of their existence. Personnel stability over a sustained period of time can help rebuild that loyalty.

A weakness notable in many commodity company cultures is an inability to capitalize on the *value* of product differences where they exist. Instead, they view product differences as a sales tool, whereby they can obtain or hold business at the same price as the competitor, instead of using the situation to charge a *premium*. Often, this results in an inability to pay for R&D above product maintenance levels. This weakness obviously works against the development of specialty products. If the management of a commodity culture company wishes to diversify into specialties, it would be well advised to set up such a business as a relatively independent entity or risk almost certain failure. In some companies, this approach is known as "intrapreneuring."

4.1.4 Technology Culture

Technology cultures flourish in R&D organizations, such as many start-up specialty polymer companies. In larger firms, they are often a subculture within the organization, unless they are fused with the entrepreneurial culture present in a high-tech start-up company that has expanded past the start-up phase. There are some parallels between industrial technology cultures and the cultures found in academic science and engineering departments, where work is often performed by small teams of professionals rather than by individuals.

Technology cultures are, by definition, more focused on scientific and engineering development than on manufacturing or marketing. In some businesses, this is not unremarkable, since these latter functions usually follow, rather than lead, the development of unique or dominant technology. Since technology cultures are often associated with entrepreneurial cultures, they are seldom found in older, larger companies within the plastics industry.

Surprisingly, technology cultures are not often found within the plastics industry, but they certainly do exist. Some European polymer producers exhibit at least some elements of this culture and it may also be observed in a few processors, as well as some compounders, distributors, machinery, additives, and instrumentation companies.

4.1.5 Nationality/Ethnic Cultures

Without getting into the fever swamps of political correctness, suffice it to say that there are significant and important differences among the business cultures in various countries and ethnic groups. These are ignored at great risk to the success of the business. The differences center on what each culture values the most.

Perhaps the most common difference between North American and many other national business cultures is the notion of timeliness and urgency. Americans put great store on being punctual at appointments and meetings, getting right down to business discussions as soon as introductions are performed, seeking to obtain immediate agreement on negotiating points, and implementing plans as expeditiously as possible. In many other cultures, punctuality holds no particular virtue and is honored more in the breach than in the observance. These cultures put great store on building trust before undertaking anything else. They want to get to know and understand a potential business associate in depth before talking about any substantive points. They are likely to find it offensive to be pushed into discussions—let alone agreements—before they feel they are ready to trust the other party. They also often wish to revisit plans several times before implementation. In other words, these cultures place greater value on building long-term relationships than they do on getting things done *now*! All of this may be very frustrating to Americans, who had better develop the patience and understanding necessary to accommodate these views or give up trying to do business where these cultures prevail. Often, turbulent periods in the history of these regions underlie the reasons why their business cultures have evolved in this direction.

The importance of trustworthy relationships in some cultures also carries over into the area of problem solving. Americans need to be especially careful when analyzing reasons for failure or inability to achieve goals when working with other nationals, and to deal with them as sensitively and objectively as possible—generally more so than when dealing with other Americans. If your customer, supplier, or employee from overseas thinks that you are engaged in a blame-fixing exercise rather than finding a solution, then they will dig in their heels and resist cooperating. American lawsuits are derided and feared almost everywhere overseas, and other nationals are often suspicious that Americans will try to use the courts to gain what they cannot through discussions. Trust, like its twin, a good reputation, is slow to build but quickly lost.

American business culture is both admired and deplored in many countries, sometimes by the same people. When people talk about finding a "third way," for example, between American economic freedom and socialist/controlled economies, they often mean that they admire American economic results but don't want to change what they are doing to obtain those results. Americans would be well advised to avoid making a practice of directly comparing American business methods to the local ones, unless they have an uncommon ability to do so very diplomatically. It is not difficult to win an argument but lose friends and business.

Nationality and ethnic business culture clashes are most acute in such situations as foreign-owned US businesses or US-owned foreign businesses. Even if the parties are fluent in each others' languages, confusion can and does arise from cultural, rather than linguistic, differences. Bob Schulz, a friend of the author, once ran a US company that was owned by a large British firm; he characterized occasional misunderstandings as being those of "two companies separated by a common language"!

4.2 Tailoring Organizational Form to Business Needs

There is no "one size fits all" when it comes to organizational form. In fact, the individual and combined strengths of the particular professionals making up the organization is more critical to successful management than the organizational format itself, especially in small companies. Nevertheless, as companies grow in size and their business in complexity, it is sensible to take steps to ensure that the customer's needs are indeed the focus of the organization rather than the process of the business itself. Functional or geographic organizations are the most common forms, but market or product organizations are also widely used. The latter two forms are also called "business units" in some companies.

4.2.1 Organizing by Function

The traditional functional structure, shown in Figure 4.1, has department heads for manufacturing, R&D, sales & marketing, administrative services, all reporting to the CEO. This is the simplest organizational form and the one used by most companies, especially small ones. It concentrates expertise in each of the units responsible for carrying out specific duties necessary for the business to operate on a daily basis as well as in the future. It does require, however, that the CEO ensure that coordination between the units is ongoing and working satisfactorily. In some companies, particularly those who have an active acquisition program, the functional units report to a COO, with the CEO more actively engaged in planning, acquisitions, etc. In effect, the COO is responsible for *how* the company is doing—the present—and the CEO for *where* the company is going—the future.

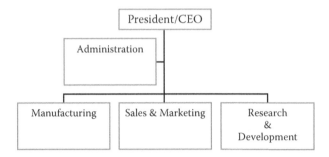

FIGURE 4.1
Functional organization.

The traditional organization can easily serve the needs of most companies, whether they are large or small, multiproduct or single product, multimarket or single market. Sometimes the functional teams may show signs of becoming too inwardly focused, and this is the time to consider reorganizing management along one of the other structures described below. Another drawback to functional organizations is that valuable but more specialized, smaller product lines and markets may not be served adequately because they become buried among major product lines and markets served by the company.

4.2.2 Organizing by Product

Many companies in the plastics industry have evolved from a functional structure to one organized by product, as they have grown larger in size and their product line become more diversified. An illustration of this structure is shown in Figure 4.2. This type of organization can take several forms. One common version has divisions devoted to a single product or to groups of products, with each division containing a functional organization. Another common form involves a centralized function for, say, manufacturing, with product units that incorporate R&D, sales, and marketing. Product organization helps to rationalize manufacturing sites and to ensure global product standards and

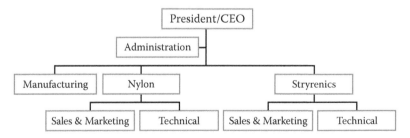

FIGURE 4.2
Product organization.

pricing policies. Not to be overlooked is the possibility of centralizing R&D on common areas where they exist. This may then permit achieving a critical mass large enough to make breakthrough technology advances, as compared to incremental improvements typical of smaller unit capabilities.

Organization by product makes sense when the company has several, relatively large, basic products that serve markets that are relatively independent of each other. For example, if a company were to make polyolefins and acrylic resins, or proprietary parts and custom parts, these products differ significantly in manufacturing terms as well as the markets into which they are sold. This would be a situation where it makes sense to have a division based on each product, with each division containing its own functional units.

4.2.3 Organizing by Market

If a company is to be truly "market focused," then organizing by market is often the most effective tool to do so. Figure 4.3 depicts a company organized by market. This structure works best where the company has a number of products whose end-use markets overlap each other, or there are a limited number of customers but whose needs can be served by a number of products made by the company. Market units that are directed to the automotive and electronics industries are particularly common. The focal point of a market-organized company is on solving application problems at one company that can be applied to other companies. This approach is particularly fruitful in such industries as electrical/electronic, where the emphasis is on participating in new applications that are likely to be small volume individually but large volume in the aggregate.

Market-focused companies are (or should be) customer-focused companies, particularly if the market is made up of a small number of customers, such as the major automotive companies. The principal disadvantage

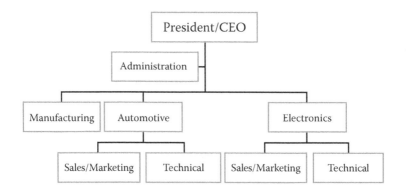

FIGURE 4.3
Market organization.

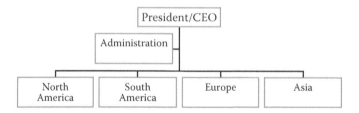

FIGURE 4.4
Geographic organization.

of organization by market is that it tends to shut out or overlook potential business in nontargeted markets for the company's products.

4.2.4 Organizing by Geography

Organizing by geographic areas is useful when speed of delivery is critical, such as for customers who demand "just-in-time" deliveries. It also speeds up decision making when there are multiple time zones between headquarters and the local plant. An example of this organizational form is shown in Figure 4.4.

This organizational form, or some variation of it, is virtually essential for companies with multinational locations. While overall policy may be set at corporate headquarters, there must be adequate downward delegation to overseas sites to handle business decisions locally, or the company will find itself always reacting to competitors rather than meeting them on at least even terms. Geographic organization delegates authority to the local managers for pricing flexibility and the ability to tailor products for local needs. Naturally, this requires that local managers will be held accountable for local profitability, too.

A serious disadvantage of a geographic organization is that it may lead to different company units competing against each other, with attendant profit erosion as units fight over business that was yours to begin with. Other disadvantages are that it leads to a certain amount of duplication, particularly R&D, and that senior management will need to be involved in coordination, to ensure that global (or even national) standards are being followed.

4.2.5 Hybrid Organizations

One can have a hybrid organization, particularly in large companies, with different groups structured by function, product, market, geography, or even by a few large customers. Figure 4.5 illustrates a possible hybrid company organization. Hybrid organizations can avoid the weakness inherent in a "one size fits all" approach, be it function, product, market, or geography oriented—there really is no single "best way"—and a number of companies find this a flexible solution when the other routes don't work efficiently. Management should always look for the most efficient and effective structure

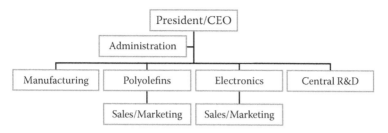

FIGURE 4.5
Hybrid organization.

that fits the nature of each principal line of business. Hybrids can be a challenge for management to ensure that they continue to work to the company's advantage; in the example shown, it would be wise to periodically check that the manufacturing and central R&D units are serving the other two sales and marketing units satisfactorily.

4.3 People Management

Elsewhere, the point is made that a command-and-control style of management is usually unsuitable for the plastics industry—or any other that depends heavily on creative workers, especially technical ones. Peter Drucker was one of the first academics to recognize this phenomenon, dubbing these individuals *knowledge workers*. It is not possible to command creativity, in any function within the company, be it R&D, marketing, sales, manufacturing, or administration. Creative solutions to problems arise from people recognizing and sharing pertinent information from many sources. Creativity is greatly diminished if information is too compartmentalized and dissemination too limited. Committees and project teams are an important way to ensure that needed information is shared and communicated to the people who need it.

Management, particularly senior management, needs to know what is going on within the company in order to execute their duties effectively. One way is to form an executive committee, consisting of department heads, and holding weekly hour-long briefing meetings—a classic, well-grounded way to keep informed. Don't simply accept reports; always ask questions! These meetings are an important way to build a management team relationship by developing an appreciation among subordinates of what each one does and how the pieces all fit together. You can further improve the quality of information shared in this way and build morale by having each of your subordinates bring along one of *their* subordinates to give a briefing on what is happening in their area of responsibility. These one-level-down individuals should be rotated from week to week so that all of them have a chance to participate over time.

Other ways to develop a better understanding of your organization's workings include being an ex officio member of every company committees and attending meetings on an irregular, unannounced basis. Your participation should not be for the purpose of running these meetings, but to let you listen and learn what's happening, how matters are being handled, who is doing the handling, and how well. If you approve, be sure to say so. If matters are not being handled to your liking, talk to the individuals involved privately and outside the meeting. This follows the old maxim, "praise in public, correct in private."

You can also practice "management by walking around ("MBWA")," a term that originated in the Peters and Waterman 1982 book *In Search of Excellence*. The idea is for top-level managers to make periodic, informal, unannounced visits in the company's workplaces, to observe and briefly talk to individual workers to find out what they think of their jobs and the company. The purpose is *not* to bypass supervision or to tell people how to do their jobs, but to *listen and learn* about how your company is functioning. You may be surprised by what you may learn this way when the information comes to you directly and unfiltered through layers of subordinates. It also helps employee morale to see the boss taking an interest in even entry-level employees and listening to their views about their work. Make sure that you inform their boss about any potentially worthwhile suggestions you hear, praise both employee and boss for their interest, and follow up later to see if the ideas have been implemented and with what results.

Be cautious about using an "open-door policy," however. Depending on how it is implemented, it can be an effective safety valve—or a major waste of time and potentially damaging to organizational relationships. There is nothing inherently wrong with being willing to meet with anyone from any level in the organization, up to a point! First of all, you should not allow individuals to invite themselves—they should go through their supervisors and managers in a direct line to obtain an appointment to see you—but their bosses should not be able to deny them the opportunity arbitrarily. While the individual employee's supervisors don't have to be present during the visit, they will resent being bypassed or not informed that their subordinate has been in to see you without their prior knowledge and consent. Second, you will need to put a limit on the number of such visits you will allow, or you will find that you don't have enough time to perform your primary duties.

An important exception to these visit guidelines would be where an employee wants to "blow the whistle" on wrongdoing (which may involve a supervisor or manager). Clearly, the employee must be able to bypass the system and see you immediately under such circumstances. Nevertheless, it is important to realize that there is more than one type of whistleblower. The first type genuinely wants to work in an ethical and honest company and is willing to put their job and reputations in jeopardy, trusting that telling you about serious wrongdoing will lead to things being put right. The second type is someone who deliberately makes trouble for others, stemming from either personal animosity, or a personality disorder that compels them to lie and make trouble for others. Needless to say, you must sort out immediately and accurately which type you are dealing with, by determining the truth or falsity of their complaint. Next, you will need to take appropriate action without delay. You will find it important—if not essential—to use a law firm that is experienced in such matters, not only to determine the truth of the accusation, but to help you deal with the consequences, including lawsuits and even government actions.

Potential lawsuits include any initiated by a false whistleblower against you and your firm, wrongdoers in your company who have been revealed and dismissed, and governmental authorities for violations of local, state, or federal laws. As long as you can demonstrate that you dealt with problems when they were detected, immediately and effectively, and have been alert to perceive such problems before they become serious, you should be able to mount an effective defense. Make sure you include your "good" whistleblower in your legal defense plans, too!

4.4 The Board of Directors

The corporation is by far the leading legal form of plastics industry companies. This is because the corporation format sharply limits the exposure of the owners and executives to individual legal liability. This is clearly superior to partnerships or sole proprietorships, which are much simpler and less expensive to maintain but lack such protection.

The governance of a corporation is vested in a board of directors at its pinnacle. The board represents the stockholders, who elect the directors at regular intervals, usually every year. The board meets regularly with corporate officers, usually the company's chief executive and chief financial officer, to receive and approve compensation, business plans, reports, and budgets, changes in the company's principal lines of business, mergers, acquisitions, etc. The board will elect a chair (this may be the company's CEO, but it is the author's view that it is better if an outside director fills this position, relieving the CEO of board administrative work). Depending on the company size and level of activity, the board may meet as frequently as monthly, but must meet at least annually.

While it is not uncommon for directors to be friends or subordinates of the CEO, this is not usually the case in large, publicly held firms. Typical boards are composed of independent directors, usually current or retired CEOs of other companies, usually with experience in similar lines of business. Boards sometimes include attorneys and bankers with expertise in the industry. Not really desirable but occasionally observed are board members who appear to have been chosen for political connections or links to special interest groups. Directors should have at least some minimum stock ownership in the company, preferably acquired before they are elected. Directors, serving on such key committees as compensation and audit, should never be company employees. It has become standard practice for directors to limit their board memberships to not more than three companies, none of which are direct competitors, customers, or suppliers of each other to any significant degree. In the author's opinion, even three boards are too many—two would be a more reasonable number of boards on which to serve, in terms of being able to genuinely contribute.

Directors should be compensated in proportion to the time they are required to serve in meetings on company business. This is often accomplished by paying a retainer plus a fee for each meeting attended; the board chair and committee chairs often receive an additional retainer. In some small companies, director compensation is paid in the form of stock options, thereby avoiding direct cash outlays.

The point of the foregoing discussion is not to instruct publicly owned companies on the need for a board of directors and how it will operate; it is to inform small, privately held company owners on how a board would benefit their business operations. Too often, small, privately held companies have inactive boards that consist of family members that may only meet on paper once a year to satisfy statutory requirements of the states in which they are registered. This is unfortunate because the owners of such companies are the very ones who would be the most likely to benefit from having an independent board that will give a fresh point of view on how the company can grow and improve its return on invested capital. For a relatively small cost, a small business owner can have between two and four outside members who can bring valuable industry experience to the company when reviewing current operations and finances, and help plan for future growth.

Should a company become involved in an overseas joint venture or have a wholly owned subsidiary, these entities will have to have a board where you will need some local representation. Your local attorney is usually the most cost-effective choice to represent you at statutory meetings; you and one or more designated subordinates will need to attend meetings where corporate strategy and budgets are on the agenda. Whoever the CEO is, this individual usually has the job title of managing director and serves as board chair.

A unique exception to the corporate governance described above is found in the Federal Republic of Germany, as a result of its "co-determination" laws, which require employee union representation and input. A supervisory board (*aufsichsrat*) consists of two groups of equal numbers of representatives, one group elected by the stockholders and one group elected by the employees. The supervisory board, in turn, elects the members (who are titled executive directors) of the management board (*vorstand*). The management board, in turn, develops business plans and budgets, which are subject to formal approval by the supervisory board. The executive directors are typically department or division heads of various business or functional units of the company. The management board is headed by a chair, who has no specific operational responsibilities other than corporate administrative and financial functions. While the individual who holds this post is not a CEO (since no such position exists), his or her decisions are almost always accepted by the rest of the board members as *primus inter pares*. With this exception, the German legally mandated management board is analogous to the executive committee used by many North American corporations.

During the author's employment as managing director of BASF's North American Engineering Plastics business unit (but reporting to an executive director in Germany), he had occasion to observe these uniquely German organizations in action, but did not note any great differences in outcome compared to US organizational practices. This form of organization seems to be unique to post-World War II Germany, reflecting a desire to allow organized labor a voice in the business practices of corporations, as well as a distrust of authority that is concentrated and not subject to checks and balances. It also appears that another goal of this particular organization form is to have a more measured and considered response to change than would be the case otherwise.

5

Managing for Success

In Chapter 1, the topic of setting goals and developing business plans was broached. Part of drawing up and carrying out business plans is utilizing both the formal and informal elements of your organization and personnel.

5.1 Planning for Success

Business plans should not be lengthy or they run the risk of becoming inflexible, or equally bad, outdated. The trouble with overly comprehensive plans is that it impossible to forecast business conditions for a year or more with any great accuracy or certainty. Therefore, it is essential to have plans that cover the range of likely possibilities, from high to low. This allows the company to be prepared for changes from the original forecasts. The plastics industry has been subject to increasing volatility in demand during the past three decades, and there is no reason to expect that this condition will improve in the future. Changes in demand are almost immediately felt by suppliers due to the increased use of supply chain management techniques, which immediately transmit fluctuations systemwide. With inventory levels designed to be at low or even "just-in-time" levels, a drop in, say, automobile sales, can translate almost instantaneously into a shutoff of parts requirements, felt through the whole system—from processor to distributor to compounder to polymer producer. Likewise, a sudden surge in demand puts pressure up and down the entire supply chain. It pays to have sufficient rapport with your customers that you can come to some agreement with them to provide for some cushioning from rapid changes.

Demand volatility is only one source of deviation from forecast or target revenues. New product development can be thrown off track by unexpected difficulties in scale-up. Your customer can be acquired by another company that decides to scale back or even discontinue the line of business on which you were counting. An unexpected breakthrough at a customer may suddenly ramp up the demand for one of your products. You need to have at least considered what you might do should any of these events take place, requiring that you react in a timely way. A range of plans—you don't need to be too detailed—will answer this need and allow you to move into action.

Plans are not merely a collection of goals and the considered tactics to reach them. They must include the resources required to execute them, both human and hardware. As described in Chapter 7, you also need to assess how you will utilize those resources by their *quality*; the best must be assigned to the most productive projects.

5.2 Managing and Integrating Functions

In order to achieve sustained earnings growth, successful managers must learn to *integrate* the direction of functional groups, not *centralize* their direction (as is often and erroneously assumed). Centralized management puts all decision-making authority in one place; integrated management distributes decision-making authority to the lowest level capable of handling it, but ensures that all decisions are coordinated toward achieving a set of common goals.

As touched upon in Chapter 1, a manager needs to be aware that the majority of the key people he or she is directing are "knowledge workers," a term coined by Professor Peter Drucker. Knowledge workers must be managed differently than others. Their contributions come from their own creativity. This requires using an open sharing of information and decision making, not a command-and-control style of management.

If functional groups do not closely intermesh and coordinate their activities, costs will rise and customers may be lost. Management must never allow "empire building" or "turf wars" to take place between functions. These kinds of negative activities are focused inward on personal objectives instead of outward on meeting customers' needs. Management needs to take firm action to prevent them from starting and drastic action to stop them if somehow they have taken place. Make it clear to your subordinates that their appraisals depend on how they are contributing to *company and customer* objectives, not on winning some sort of perceived internal competition. During the Jack Welch years at General Electric, much was made of using internal competition, for example, peer versus peer, in order to push growth faster. However, this appears to have been rarely practiced elsewhere; examples are lacking of other companies that made this idea work successfully on any sustained basis. As you may infer from the views expressed by the author so far on managing subordinates, internal competition makes little sense as a way to motivate people to cooperate with each other.

The use of project teams and committees is central to integrated management. The input of each affected function is thus made part of the solution to dealing with problems and improving operations. Committees do much of the coordination of the various functional groups in a plastics industry company without requiring that senior management be directly involved. Standing committees are needed to handle matters that are routinely repetitive, for example, raw materials qualification and purchasing. In the latter instance, R&D can present the formulation characteristics of the materials under consideration, manufacturing can present the processing characteristics, marketing the customers' preferences, and purchasing cost and logistics considerations.

Project teams are committees that have a specific, short-term purpose, and disband once this purpose has been accomplished. They include members of different functional groups or different engineering disciplines. An example might include the development of a major new product for a large customer or even just putting together a logistics system that will coordinate the business requirements of the company's largest customers. When a project team has achieved its objectives, then disband the team. If the team is not achieving required milestones, reconsider its objectives and the resources assigned and make changes where needed. Sometimes you may find that a project team works so well together that important synergy would be lost if it were disbanded. In such a case, consider restructuring the group to be a committee with newly assigned objectives.

A word of caution: create committees sparingly, keep their membership size limited (six would be the maximum), and insist that their meetings be short (less than an hour) and never more frequent than weekly. Information can be shared through e-mail and memos, but discussion works best when everyone can discuss issues face to face. Poorly run committees are a terrible waste of valuable time and damage morale. Well-run committees keep everyone "in the loop," maximize efficiency, and minimize mistakes by building teamwork. The secret to successful meetings is for the chairman or chairwoman to stick closely to an agenda that addresses key issues only, getting agreement on work assignment scope and milestones before adjournment, and handling issues with noncontributing members outside the committee meetings. Spread committee work around, too. It's good for members to gain experience by rotating membership among different people in the same functional group; it also keeps diversion from group members' primary tasks to a minimum. It is a telling sign that "committee-itis" has set in if customers, suppliers, or other employees find they cannot contact people who are members of committees because they are always tied up in meetings.

5.2.1 Research & Development

R&D is fundamental to the success of any company in the polymer manufacturing, compounding, processing, or equipment/additives sectors of the industry. Only distribution has little need for R&D, and this can normally be handled via consultants. New products and processes are the lifeblood of sales and earnings growth in a competitive marketplace. However, this does not mean that R&D has *carte blanche* to do whatever takes its fancy. Every new product that looks promising in the laboratory must be vetted by marketing that a potential market exists and by manufacturing that it can be produced at acceptable quality and cost levels.

As a bad example, take the real-life case where a large commodity polymer producer's R&D developed a new engineering polymer that exhibited an excellent balance of properties. Without obtaining more than minimal market research or going through the full range of production scale-up steps, the company committed to a commercial-scale plant. The new product was sampled to customers and initial orders obtained. Then some applications began to report field failures and several molders complained that the product showed variable lot-to-lot processing characteristics. Manufacturing found that the plant design would not allow the product specifications to be met at anticipated costs. After 18 months of trying to put out these fires (relatively unsuccessfully), management decided to divest the business, putting the plant and product line up for sale but found essentially no takers at anything other than a token price. Eventually the plant was closed and the entire project written off, at a hard-to-believe cost on the order of magnitude of several *hundreds of millions* of dollars. The manager in charge of the project was reassigned, but lower-level personnel were dismissed as part of the restructuring. Admittedly this is an extreme case, but it happened in a company with seasoned management that made the astonishing blunder of expecting the scale-up process and costs for a developmental specialty to be essentially no different than those for a variation of an established commodity product.

One of the most difficult but critically important jobs a manager has in the plastics industry is to successfully integrate technology and marketing. Plastics materials, even commodities, cannot be sold like detergents; potential and actual customers must know *how* to use them in order to gain the benefits of their features and be able to pay for the value they confer on an application. The speed and relative success of this learning process, for example, "time to market," can make or break the profitability and competitive advantage of a new product for a company. The most successful companies in the plastics industry combine their product development and marketing groups into teams. This ensures that the company's technical resources are used to meet customer needs with the least amount of "filtering" in communications and time loss.

At one time, most polymer producers offered design services to qualified customers as part of their sales and marketing "package," to stimulate and encourage new business opportunities. Computer-aided-design (CAD) and computer-aided-engineering (CAE) services can be an important benefit to prospective customers by reducing time to market and risk. To the best of the author's knowledge, this is a service no longer offered by most polymer producers. However, many compounders and processors have such capabilities, but these services are normally offered for a fee. In the author's experience, this is an important, if not essential, component of business development. If the product design of a new application is sufficiently unique, it is likely to be worthwhile to apply for a patent, which can be used to cement the customer relationship more tightly.

5.2.2 Sales and Marketing

It's been said by some that marketing savvy seems to be in short supply in this industry while sales know-how seems to be in abundance; only a few companies seem to know how both of these functions work. Everyone knows what the term "sales" means: getting orders from specified customers. The meaning of "marketing" seems to be less well defined. In the author's opinion, the problem is basically a lack of understanding of what marketing really does, how to use it, and how to integrate marketing with sales. For one thing, let's clear up one point now: marketing is *not* a series of advertising campaigns or a blitz of press releases, nor is it primarily needed to find and develop new customers. Marketing is much more than this. Marketing includes the following:

- Identifying, analyzing, and aggregating a series of individual similar customers and applications into markets. For example, a transparent polymer such as polycarbonate that is sold to customers for optical applications—that may be further classified as for consumer and automotive end uses.
- Identifying and projecting trends in markets to help the company anticipate and manage the coming changes. Example—the commoditization of most electronic devices has lead to a decline in profit margins and increased offshore sourcing.
- Transforming recognition of the company's products, services, and conduct of its business into respect and loyalty among its existing customers, and then finding ways to bring this recognition to the attention of potential new customers.
- Collecting, analyzing, and transmitting the information necessary for the company's future business plans.
- Sales and marketing are complementary, not competing groups.

No one seems to have a problem knowing what salespeople have to do, but sometimes they do seem to have a problem understanding *how* they have to do it. Sales work does not consist of a big entertainment budget; if purchasing agents were so easily seduced, their bosses would notice quickly that they were not doing their job of buying the best for the least. A good sales representative will identify *all* of the customer's important decision makers and make sure that he or she communicates with them on a regular basis. It is not enough to know that the company is meeting the current needs of the customer. One must also know what their future needs will be: whether there will be more or less business to seek, what the customer's financial situation and business goals are, etc. This information is needed not just to serve the customer, but for future planning as well.

A critical part of both sales and marketing is determining pricing for your products. When you are targeting new applications, your pricing should be primarily based on the *value* your product brings to the user. Your best chance to capture this value is early on—competition will eventually bring pricing down, but if you cannot price at a premium initially, then you need to examine the reason very carefully and make sure that you take corrective action to better position the company's products and services in the future.

5.2.3 Manufacturing

Manufacturing is all too often taken for granted, but its consistent, quality-conscious, timely, and cost-effective execution is crucial to the success of any nonservice company in the industry. The leading firms of the industry have adopted Total Quality Management (TQM), Six Sigma, or other such techniques to ensure continuous improvement in consistent quality, which also typically leads to cost savings.

Manufacturing consists of processing raw materials into finished products. Your suppliers of raw materials plus the logistics companies that transport and store these materials constitute your supply chain. You are only doing half the job if you are managing only your own production scheduling and not the rest of the supply chain. It is impossible to control costs and quality if you have not brought your suppliers on board as partners. ISO 9000 certification for your company and your suppliers is an integral part of ensuring globally consistent manufacturing quality, just as much as are Total Quality Management and Six Sigma programs.

There are a number of software programs that assist manufacturing management, generically known as enterprise resource planning (ERP). These systems allow a company to employ a relational database globally that keeps track of all its purchases, inventory, manufacturing scheduling, and shipments. Supplier-/customer-compatible ERP systems can effectively integrate the production process of both, with significant potential savings and short-ened lead times. Early ERP supply chain systems were the forerunner of much of what is called e-commerce today. In addition, a properly enhanced ERP system enables management to analyze its return on investment in a variety of ways.

The installation and training costs for these systems are substantial. The time required to train personnel in their usage normally varies with their familiarity with information technology (IT) systems; six months to a year is not unusual. While the ERP suppliers will gladly supply a package deal, it pays to shop for less expensive alternatives, even unbundled ones, which can result in substantial savings. One method for keeping these costs down and accelerating the system's implementation is to use one of the standard templates offered and avoid customization unless your requirements are justifiably inflexible. Customization costs more, takes longer, and raises the risk of running into "bugs" down the road. All of these systems need annual maintenance, as would virtually any asset.

Still smaller, one-site systems are available for processors at proposition-ally lower costs. These smaller systems are focused on manufacturing and therefore may lack some of the sophistication of larger systems, but they are easier to install and use. They are a good way to become familiar with the concept of ERP instead of jumping in with an accompanying large invest-ment of time and money (and risk).

The established use of e-commerce is testimony to the efficiency of using either private networks or the Internet for exchanging order and inventory information between suppliers and customers. The technology involved is now relatively established, as is the security of transactions. This all falls under the term "supply chain management (SCM)." The rise in the use of SCM has been accompanied by buyers pushing to reduce the number of their suppliers. Fewer suppliers usually mean more purchasing leverage both in terms of lower prices and more services included in the price. If practiced correctly, SCM also offers lower transaction costs for both parties. Fewer sup-pliers should find that the closer relationship with the customer means a greater ability to provide more products and services than before, and to participate jointly in new developments.

There is a distinct risk to a reduced supplier base, however. A smaller sup-plier base can expose buyers to delivery delays and product quality prob-lems if they lack the ability to turn immediately to other suppliers to fill in the gap. Also, underfinanced suppliers might have problems obtaining raw materials soon enough to meet an unexpected order from you.

The ultimate "reduced" supplier base is to source from just one com-pany. Besides the problems mentioned above, perhaps the most serious drawback to a single source is that you are likely to cut yourself off from new products and technologies developed by suppliers with whom you no longer have a relationship. As an example, a parts division of one of Detroit's "Big Three" automobile manufacturers had an exclusive sup-ply contract with a major polymer producer in the United States. A European competitor of this producer called on this division with a pro-posal to reveal a new technology for making large parts at a lower cost, in return for a second source position. The parts division refused, citing their exclusive contract and then asked their supplier to furnish the same technology to them. The domestic producer took 18 months to do so. In the meantime, the European competitor had gone to another Big Three firm and had their proposal accepted almost immediately. The estimated savings to the smarter Big Three company that took up the European proposal ran into millions of dollars and helped them to increase their market share.

5.2.4 Administration

Administration is a sort of afterthought for some, but it has an important impact on a company's performance. Administration has a number of components:

- *Human resources* (HR) must serve management's needs to find, hire, and help retain top-quality personnel. HR has to maintain current information on industry-wide compensation and benefit practices, as well as keep up with frequent regulatory changes in this area. HR must also administer the performance review system to ensure that it is running on time and properly.

- *Finance* must manage the company's cash flow so that the company collects and disburses on a timely basis, at minimum net cost. This responsibility also includes maintaining lines of credit at banks and monitoring stock and bond markets for suitable opportunities to raise money if the company is publicly held. Such activities must be conducted with the utmost integrity and transparency; it is insufficient to be merely "legal."

- *Management information services* (MIS) unit must provide necessary financial and operations data, accurately, and on a timely basis, to each level of management that needs these data to perform their duties properly.

- *Credit* must correctly and continuously assess the ability of customers to pay fully and on time, and communicate this information to management promptly. A *few* bad debts may an acceptable risk associated with aggressive selling, and no bad debts may indicate too cautious an approach, but many bad debts can injure the company's financial performance, which is unacceptable. Credit insurance may be a reasonable option if management is uncomfortable with the credit worthiness of some larger but riskier customers.

- *Legal* must ensure that the company's contracts protect the company's interests, defends against lawsuits brought against the company, and initiates lawsuits where the company has been injured by another party and a fair settlement cannot be negotiated. Legal will find outside attorneys to handle matters that require specialized knowledge, for example, antitrust and patents. To obtain the best results from the legal department, tell then the outcome you believe is fair to both parties and ask for advice on how to best obtain this outcome. Remember that anyone can sue anyone else at almost any time for almost any reason in the United States! Nevertheless, you need to strike a proper balance in legal matters, otherwise you will find yourself refusing to take even reasonable risks and missing opportunities.

5.3 Managing Costs

One of the most challenging aspects of a manager's job is properly controlling costs, particularly during economic recessions. The old cliché about "doing more with less" never seems to go out of style, regardless of whether the overall business climate is growing, slowing, or stagnant. The press seems to cover news only about plant closures and layoffs. Managers come under pressure to show that they, too, can make tough decisions about cutting staff and closing plants. The problem is that these are not tough decisions; these are the easy ones. If "everyone is doing it," then directing these contractions will likely be the line of least resistance, allowing one to avoid criticism. It may well be that it makes very good business sense to close an old plant that cannot be economically modernized, particularly if the work can be transferred to a more modern facility where the added throughput will create overall savings. It may well be that it makes sense to close or divest a part of your business that is not growing and is only marginally profitable. It may well be that you can restructure your organization to "flatten the pyramid" by consolidating overlapping positions and laying off truly marginal performers. It may well be that it is time to require slow-pay customers to pay more or pay in advance. The point is to avoid overdoing all of these things—which you should have been doing earlier as part of good management practice! Too often, managers who feel under pressure during hard times follow the crowd or even overreact, rather than concentrate only on what needs to be done in their own companies and their specific circumstances.

A further caution about trying to restore or improve profitability during hard times is in order: Make R&D cuts the *lowest* priority. Reducing R&D expenditure is sacrificing the future for the sake of the present. Reduced R&D means reduced future new products and processes, accompanied by reduced future sales and earnings. Cutting manufacturing capacity reduces the company's ability to respond quickly to unexpected but established market opportunities, but this is often less critical than missing out on *new* market opportunities. Staff reductions, product line revisions, and business divestitures are discussed in more detail in succeeding chapters.

One area for finding cost savings that is not often considered in smaller companies is working capital reduction. If business is reasonably profitable and generating good cash flow, then management should review whether it would be feasible to reduce inventory, lengthen payables, and improve receivables. Do you regularly cull stale raw material and finished goods inventories, in order to convert it back to cash? Can your suppliers ship smaller quantities without added cost and lead-time penalties (or at least only minimal ones)? Would they be willing to allow you to pay in 45 or 60 days instead of 30, without a price increase? Can you raise prices a small amount, say 2%, on customers who are financially sound but pay in 60 days instead of 30, without losing their business to a competitor? Do not overlook spare parts kept in maintenance; they are a form of inventory and can be converted to cash, if they are not hard to replace and seldom needed. There are many things you can do to free up cash and improve earnings on invested capital if you look for these and similar opportunities.

6

Managing Globally and Sustainably

Since the first edition of this book was published, two issues have become increasingly important to the plastics industry: globalization and sustainability. Unfortunately, neither of these terms has a uniform definition, creating confusion when addressing how these subjects affect business activities. Worse, these topics have significant political overtones as well, which tends to confuse rather than clarify what is happening and what should or can be done in response. To avoid or at least minimize these problems, the author has tried to apply scientific principles and common sense to identifying what globalization and sustainability genuinely mean to the plastics industry, and what specific steps management needs to take when planning and executing business strategies affected by these two issues. Both of these subjects cannot be ignored by management, or dealt with casually, because they will certainly have a significant impact on business development and profitability over time.

6.1 What Is Globalization and What Is Its Effect?

Globalization is a relatively contemporary term commonly used to describe the pronounced increase in international commerce beginning in the late 1990s. It is often more specifically linked to the rapid industrialization of China during that period of time. The US plastics industry has been directly affected by globalization, both positively (such as new markets) and negatively (such as new competition).

In the past decade, globalization has attracted the attention of columnists in the larger metropolitan newspapers, but more often than not, these writers focus on political and sociological issues rather than economic, business management, and trade matters. Perhaps the most familiar example would be the *New York Times* columnist Thomas Friedman's *The World is Flat,* published in 2005. Unfortunately, the author's lack of familiarity with basic economic principles and business practices largely limited his work to a narrow personal perception of globalization rather than broader-based observations and analysis. The plastics industry-based reader would be better served to consult more scholarly, industry-focused, peer-reviewed books on this topic, such as Peter Spitz's *The Chemical Industry at the Millennium*; others are also listed in the "Suggested Readings" section of this volume.

Friedman claims—and he is by no means the only one to do so—that globalization has meant a continuing transfer of manufacturing jobs from the United States to offshore locations, primarily China. This brings to mind the mass media's editorial reaction to the North American Trade Agreement when it was approved in 1991: it was claimed that there would be a "giant sucking sound," as American jobs disappeared into Mexico—which never happened. When asked abut this prediction in the following years, these sane mass media editorial writers seemed to suffer collective amnesia, ignoring the question or changing the subject. While it is certainly undeniable that manufacturing jobs have been disappearing in the United States since 1980 or a bit earlier, there is no credible evidence that these jobs had moved to China, or for that matter, anywhere else overseas. On the contrary, an examination of industrial employment worldwide show that manufacturing jobs have been disappearing *globally* since the 1980s, primarily due to the application of new technologies that yield ongoing productivity improvements: the application of computers and the Internet to manufacturing supply chains being an early, widespread example.

Nevertheless, there are two important exceptions to productivity-caused job losses during this period. The first is found in the collapse of the Soviet Union in 1990, which resulted in the closure of many uneconomic state factories, with an accompanying loss of *tens of millions* of jobs. In today's Russia, there has been only a very modest recovery of jobs since that time, because the government's primary economic focus has been on expanding minerals extraction, which is a far less personnel-intensive industry than manufacturing sectors such as automobiles or construction.

The second exception is the major restructuring of the Chinese industrial model in the late 1990s, from a large number of huge, overlapping state-run enterprises to a much smaller, consolidated group of state-controlled "heavy industry" firms, for example, steel, mining, oil and gas, refineries, power generation, etc. This consolidation also resulted in the net loss of many millions of jobs; in fact, more than in any other country, including Russia. Next came the establishment of a very fast-growing "light industry" private business sector, which made up for those earlier losses in just three years, 2000–2002. For the great majority of these new businesses, products for domestic sales are every bit as important as they are for export, if not more so.

The media's intense focus on China's exports to the United States has strangely ignored the fact that total US imports from all Pacific Rim countries, including China, have actually been in a *steady decline* since 1994, according to the US Bureau of Economic Analysis. Chinese exports to the United States have indeed been on the rise and to some extent, have replaced exports to the United States from other Asian countries, but not by nearly enough to keep the total overall volume from decreasing steadily.

Another major misperception about globalization has to do with trade between the United States and China, which is often described as one-sided, with all of the advantage going to China. For some reason, one never reads that China is the United States' *biggest export customer*. Very often, US-made components are sent to China for assembly; the finished goods are then exported back to the United States—but government record-keeping counts these return shipments as being of *100% Chinese origin*. In other words, much of US–Chinese trade represents partial processing in each country to make finished products, but the US government's oversimplified reporting completely obscures this fact.

The confusion over who is selling more to whom is at the root of the questionable complaint that China has a substantial advantage over the United States in the balance of trade. While it is true that the total value of China's imports from the United States is something less than the total value of China's exports to the United States, this single measurement is not terribly meaningful because countries no longer settle their balance of trade deficits by shipping gold back and forth to compensate. If we did, perhaps the critics would have a point that we would run out of gold eventually (unless we mined more!). However, the Chinese government has been primarily neutralizing its dollar trade imbalance with the United States by buying US Treasury bonds, which has pleased the US government very much. China has been effectively financing a very substantial portion of US government deficit spending for more than a decade.

In mid-2012, China owned 22% of total US Treasury debt—which is down substantially from 37% in mid-2011, but still more than any other country— with Japan second. It is noteworthy that the press has long forgotten that it was touting *Japan's* exports to the United States as a huge threat to our economy in the early 1990s. Despite this situation being virtually identical with the current Chinese "problem," Japanese exports to the United States never produced any measurable adverse consequences. Japan also neutralized its dollar trade imbalance by buying US Treasury bonds, again with no complaints from the press.

During the Great Recession that began in mid-2008, US imports have declined but US Treasury debt has soared, due to repeated, enormous federal "stimulus" spending programs. These actions have guaranteed a decline in the purchasing power of the dollar, and China has been feeling this in the form of rising costs of imports, such as oil and other commodities. Essentially, China has been importing US inflation. China has dealt with the downturn in exports to the United States (due to the Great Recession cutting US demand) by placing even greater emphasis on developing its domestic market to continue its economic expansion, although this has not been sufficient to replace all of the lost exports. Consequently, China's GDP growth during the Great Recession has been substantially reduced. In addition to purchasing US government debt, China has been using its hoard of US dollars to buy rights to oil reserves in friendly third world countries, and acquiring business enterprises in the United States, and in other countries that will accept payment in US dollars. Less reassuringly, it is also using its dollars to expand and modernize its military forces, which are second only to the United States in size and capability.

The Chinese government has also been raising the exchange rate of its currency, during the course of the Great Recession, at a more rapid rate than previously. In 2009–2010, it strengthened the yuan versus the US dollar by nearly 10%, more than double the rate of increase in the four years since the yuan was floated in 2005. Clearly, the Chinese have been doing this in order to reduce the amount of imported price inflation mentioned earlier.

Several years before the Great Recession began, the United States protested China's currency policies to the WTO (World Trade Organization); the US government took the position that the yuan was deliberately undervalued versus the dollar in order to create an unfair competitive advantage. However, the WTO determined that the Chinese actions were within the WTO rules, and rejected the complaint. Nevertheless, many US politicians, as well as metropolitan area news media, have continued to assert that the Chinese are using currency manipulation to steal jobs from the United States, while ignoring that US consumers, especially poorer ones, are benefiting greatly from being able to buy less expensive goods. In the opinion of this author, these political criticisms are misdirected; they ignore far more important government-imposed burdens on US companies (including exporters), such as the uniquely high US corporate taxes, as well as regulatory and litigation costs that are far more onerous than those of virtually all of the United States' offshore-based competitors, including China. They also studiously ignore the "elephant in the room": the fact that the US Federal Reserve has been literally printing money at a rate greatly in excess of our economy's needs, for the purpose of "stimulating the economy," or "quantitative easing," since 2008. Whatever the term used, these activities are every bit as much a form of currency manipulation as is China's buying and selling its currency to maintain its exchange rate within a designated range. The answer to this perceived competitive shortfall is not to blame others for an uneven playing field (which the US government has played an important role in shaping), but to put US economic policies in order and stop debasing the dollar.

While China is often the main focus of globalization discussions, one cannot ignore India in this context, as it is bidding to become the "next China". The Chinese phenomenon is destined to reach its pinnacle within 20 years and then decline, just as Japan is now experiencing and for the same reason: population growth declining below replacement levels. This decline in total population is accompanied by a significant increase in the average age, and history shows clearly that aging populations produce less and consume less of everything except medical services and pharmaceuticals. Thus, the economies of such countries shrink along with their populations. Russia is well into this decline now, and many Western European countries, particularly Germany and France, also face this same problem. These population changes will strongly affect future global marketing plans for many companies.

Analysts at the Population Reference Bureau and the Central Intelligence Agency are forecasting that the population of India will be greater than that of China by the year 2032, only a generation away. This prediction has a high probability of being realized due to the "one child" policy that was adopted by China for urban married couples in 1979. India has the second largest population in the world, and an accompanying strong, expanding economy. The United States has the third largest population in the world, but by far the largest economy. Indonesia and Brazil follow in population size, but have much smaller economies. Depending on a company's product line, market emphasis, and plans for growth, India, Brazil, and Indonesia offer future sales and growth opportunities that will have different timelines than in the United States. It is a little late in the game for any US companies without a current presence in China to make a major commitment now. It would be wiser to consider other countries, especially India and Brazil, as offering potentially greater future opportunities and growth.

6.1.1 Globalization and Sectors of the Plastics Industry

Overall, the plastics industry worldwide is vibrant and growing. Nevertheless, this will vary significantly from region to region and sector to sector, depending on area market conditions and the levels of technology achieved. Accordingly, the primary emphasis in the following analyses will be from the viewpoint of US experiences, unless otherwise noted. The impact of the Great Recession and the lack of recovery to date have had the greatest influence on business conditions by far, of course. Europe has also suffered from a decline in the value of the euro versus the dollar; since the Europeans are also printing money at an unprecedented rate, it seems unlikely that the euro will recover its earlier premium against the dollar anytime soon. While this will make European exports cheaper in overseas markets, it also has the effect of lowering the value of European investments and making them possible targets for acquisition by offshore companies. This has exposed the danger of the EU's Faustian bargain: having a common currency that lacks common financial controls. The southern countries, particularly Greece and Spain, have seen unemployment rates soar well into double digits, as years of heavy deficit spending by their governments are now causing a brutal contraction of their national economies; Italy and Portugal are not much better off. It is difficult to predict what will happen next, as leaders of the European governments have been unable to agree on any realistic solutions. One scenario would be for Greece and other similarly troubled EU members to abandon the euro and return to their previous national currencies, which they will devalue in order to pay off debt and make a fresh start. Thus, the economic outlook for much of Europe is not promising in the years to come.

6.1.1.1 Machinery

Chinese and European competition has been strong in plastics processing machinery, a sector in the United States that was virtually crushed during the two recessions of the past decade. US custom processors were also hit hard in these downturns, and this has been likely the primary reason for the machinery sector's decline: the liquidation of bankrupt processors flooded the US-used machinery market in the recessions of 2001 and 2008. Indications are that this glut is finally being eliminated, and surviving machinery makers are again seeing demand growth.

Some advances in processing technology are also helping restore growth, such as equipment that allows glass fiber to be added directly to molten polymer and then molded, eliminating a separate compounding step. This has so far been mainly utilized in large automotive parts, such as front-end modules, which has also limited equipment sales with this feature to a relatively small number of large molding presses.

6.1.1.2 Processing

The winnowing out of custom molders, mentioned earlier, has resulted in the reduction of as much as one-third of the US firms formerly in this sector through consolidation and bankruptcy over the past decade. The surviving firms are leaner, more focused, offer more services, and often have cooperative arrangements with offshore molders (if not outright ownership). It goes without saying that success for any company in this sector will lie in building a business that uses increasingly sophisticated equipment to improve productivity. Customers also value technical service and custom designs very highly, which need not necessarily be in-house; a good working relationship with one or more outside firms is often a sound solution. Bank credit is likely to remain tight for some years to come and this will make retained earnings a more important source of capital than borrowing for some time to come.

6.1.1.3 Polymer Manufacturing

US polyolefin producers were also brought low by the Great Recession, combined with high natural gas prices and the start-up of several Persian Gulf polyolefin plants that finally came on stream after lengthy delays during preceding years. However, beginning in 2010, the US natural gas industry began a vigorous expansion, due to new extraction techniques ("fracking"). The resulting rapid increase in gas supplies has driven prices down, and domestic polyethylene manufacturers are again very competitive globally. Styrenic polymers, however, are in a difficult position due to the high cost of aromatics and some firms, such as Dow and BASF, have divested these operations or shut them down. Engineering plastics appear to have bounced back, despite a significant run-up in costs. The latter have been successfully passed on to users so far. "Green" polymers are also beginning to grow, but there is a very real question as to how "sustainable" this growth may really be in view of their high costs, unless they continue to be favored by government subsidies and regulations.

6.1.1.4 Compounding

Interestingly, the specialty compounding sector has been less affected by the recession than others. This has much to do with the intense customer focus of these firms, particularly the smaller and more nimble ones. They have been shifting away from relatively undifferentiated mass markets, such as high-volume applications in automotive and construction, and toward smaller but greater inherent-value specialties, such as sporting goods and medical applications. Another, newer, characteristic of well-managed firms in the sector is to offer fee-based application development services rather than compete for already established applications against other compounders, who may be larger and better able to use pricing to compete. Development services also carry an advantage in that personal contact with customers is very important and can be a critical advantage over offshore competitors.

6.1.1.5 Summary

A quick examination of these situations readily shows that competition from the developing economies in countries such as China, India, Brazil, and elsewhere did not leap full-blown into existence overnight. These competitors have been building their businesses over a long period of time, using a base of imported technology combined with local low cost labor and capital. To some extent, these countries—except China—have been able to finance much of their expansion with foreign aid from the United States and UN agencies (which themselves are heavily underwritten by the United States). The local economies in these countries are growing faster as employment increases, with an accompanying improvement in living standards.

In order to compete successfully on a global basis, US and European companies must move up the technology curve and use less—but more skilled—labor, leveraged by continuously improving productivity. A simple, long-established example would be the custom molder who has installed robots to remove and de-gate parts from a mold, stacking and packing them as well, so that a single operator can oversee production from multiple machines. Focusing on markets that demand sophisticated designs, materials, and processing is essential to competing successfully against globalized competition. It is also important to avoid competing in commodity applications unless there is a unique feature, time factor, or cost saving involved, one that cannot be easily duplicated by competitors, especially ones located offshore.

6.1.2 Strategies to Take Advantage of Globalization

6.1.2.1 Weighting Defense versus Offense

It is simply unworkable to wait until imports pop up at your customers and then react. This is a losing strategy because, by definition, you will always be a step behind the competition. One cannot ignore globalization either, because this will almost certainly result in at least some loss of existing business when customers seek the most efficient combination of component manufacturing and assembly, regardless of where such suppliers are based. Therefore, it is essential to develop and carry out a positive business strategy that takes *advantage* of globalization to widen business horizons by seeking innovative partnerships and technologies wherever they are located.

Jack Welch, General Electric's famous CEO from 1980 through 2001, proclaimed a "70-70-70" globalization strategy for his company near the end of his tenure:

- Outsource 70% of the company's workload
- Use offshore locations for 70% of the outsourcing
- 70% of the offshore outsourcing will be in China and India

While these guidelines certainly grab one's attention, they were undoubtedly exaggerated for shock effect. Nevertheless, the concept of outsourcing nearly three-quarters of a company's workload suggests that Mr. Welch was convinced that coming changes would be so numerous and frequent that it will be easier and faster to buy products and services from others rather than try to evolve rapidly in-house. Such a radical change in normal business practices also carries a high risk of losing control over quality and innovation. The next two concepts of offshoring evidently presume that overseas-based supplies of products and services invariably will be less expensive, of at least equal quality, and fully competitive in delivery times, compared with domestic sources. These are very questionable assumptions on which to bet a large chunk of your company's future. Welch's third rule makes more sense than the first two, because China and India represent huge, growing target markets in themselves, and therefore can be most competitively served by producing more products in locations where they will be *sold and serviced locally*, as well as shipped back to North America for sale. It is far from clear that GE ever pursued these initiatives on a consistent, serious basis and Welch's successor, Jeff Immelt, clearly has a very different vision of what GE should be (including the divestiture of GE Plastics). So far, Immelt's vision has failed to yield the same consistent financial results that GE produced under Welch.

A globalized business is not one that simply makes products at home and ships them overseas; nor is it one that makes or buys everything overseas and then sells them at home. These types of firms would be most accurately identified as exporters and importers. A globalized business is one that has integrated markets, customers, technology, and suppliers, as conditions required both at home and abroad. This is not a simple business model and cannot be achieved and maintained consistently without top-notch talent, technology, and management.

Another important consideration that needs elaboration is the *political* climate or climates in which your business must operate. A political climate includes such aspects as taxation, intellectual property protection, regulation, and the safety of personnel and physical assets. In the last quarter of a century, the United States has been losing ground to many other nations in terms of its own political climate for manufacturing. The United States has by far the highest corporate income taxes and regulatory costs among its nine largest trading partners, which include Japan and Germany as well as China. US-government-mandated employee benefit costs are higher than all but three of those nine partners. Tort liability costs in the United States are far higher than anywhere else in the world. The US executive branch negotiated several favorable, bilateral trade treaties with Asian and Latin American countries in the early part of the recent decade, but these were never ratified by Congress until very recently, for political, not economic, reasons. It is clear that the United States' greatest handicaps to benefiting from globalized trade are the direct result of conflicting domestic political philosophies, and not due to a lack of fundamental US business competitiveness. Perhaps future national governments will attract more business people, particularly in Congress, who will work on reducing or eliminating these internal restraints on US manufacturers' competitiveness, but this will not happen overnight. In the meantime, a sound international, globalized business program will carefully consider making and selling products wherever the company's goals are most effectively and efficiently achieved.

6.1.2.2 Things to Avoid or Beware of When Buying Globally

There is a wry (and all too often true) aphorism about buying commercial items offshore: "you can have fast, cheap, or good—pick any two." This is sound advice when doing business anywhere, but particularly overseas with unfamiliar vendors. The biggest advantage of any domestic producer over one located overseas is that shipping time is minimal, often overnight (by truck), compared to the typical two-month voyage by freighter from Asia to the United States. Of course, one can always take delivery by air freight, but this is very expensive from Asia (although much less costly when shipping *to* Asia). Consequently, a buyer will need to make tradeoffs on cost versus speed of delivery, but quality is an entirely different matter. There are very few products where quality can be compromised for speed or cost, and poor product quality will hurt your company's reputation very quickly, regardless of speed or price.

Going back to the lengthy shipping times for products made overseas, one should note that there is another pitfall involved. Too many buyers overlook the fact that this delay requires more working capital than would be the case for locally made products. Why? First of all, most overseas suppliers want payment before the goods leave their factory for the docks and ocean delivery from Asia to the United States usually takes eight to ten weeks.

Air freight from Asia to the United States costs about ten times as much as US domestic air freight, so faster delivery is not a practical way to offset the shortened payment terms even partially. Second, dealing with long delivery times usually means carrying larger inventories—and needing more working capital. When the Great Recession began, banks reduced the amount of debt they would finance, and many firms that had outsourced manufacturing to Asia had to pull back. This was not just because domestic production costs were now more competitive or lower sales meant smaller inventories. The "new normal" appears to include reduced credit lines for working capital. In summary, a globalized business has to look at a complete picture of its business and financial components when making decisions about what to offshore and what to keep at home. Simply buying parts or even fully assembled items at lower prices is at best an incomplete business strategy.

If a sector of your business is dependent on trade secrets, such as formulations or equipment design, you should be extremely cautious about exposing these to non-US suppliers who are making products for your resale or components for you to assemble in the United States. The reason is that legal remedies for the misuse or theft of trade secrets in many countries located outside North America and Europe are nonexistent or weak at best. Patents do not offer much protection in these countries, either. Suing an infringer can often cost more than any recovered damages, but not suing might lead to the invalidation of your patent, on the grounds that you had "abandoned" it by failing to enforce it against infringers. Look for guidance by a good patent attorney who knows or has access to knowledge on strategies for protecting your technology offshore.

6.1.2.3 Competing Globally through Partnerships

Buying and selling products globally is often too complex to be managed effectively by a small company. The answer can be partnering, which can take several forms. For example, one can contract with a business consultant who specializes in working for firms who wish to buy and/or sell in China, India, etc. Another answer is to join a "cluster" (a B-school buzzword for a group of smaller companies with common or overlapping interests that can be shared without triggering antitrust problems). Clusters can be made up of companies in close geographic proximity to each other, for example, industrial park "incubators" for start-ups, or simply organizations whose members share common basic business interests, but are geographically separated, for example, Manufacturers Association of Plastics Processors (MAPP). The Society of the Plastics Industry (SPI) also can help with such arrangements. In particular, incubators allow for shared accounting, legal, advertising, trade show exhibitions, and personnel services. MAPP concentrates on sharing productivity benchmarks. SPI undertakes government lobbying for the benefit of its overall members' interests. There are more benefits than those listed, but they have a common focus: finding ways to help individual member companies to share costs and improve their access to markets at home and abroad.

6.2 Managing Sustainably

Sustainability is a term that has come into popular use during the past ten years or so, but it can be very frustrating to deal with because it has many meanings. In fact, the Federal Trade Commission took note in mid-2011 that no uniform definition of this word exists. At the extreme, a "sustainable society" might be thought to mean living and working at a government-enforced subsistence level, that is, a preindustrial economy. However, a more commonsensical and useful meaning of sustainability would be "economizing," that is, *using only the minimum necessary of any resource in the course of our lives and work.* Other terms and concepts used widely and usually associated with sustainability include green, reusable, renewable, recyclable, bio-based, and biodegradable, although each of these also have other multiple, nonstandard definitions that will be discussed later.

For purposes of this book, the author has avoided any linkage of "man-made global warming/climate change" to sustainability. There are more immediate and practical concerns about conserving natural resources with respect to sustainability than those raised under the highly politicized and contentious heading of "climate change."

6.2.1 Separating Wishful Thinking from Reality with Respect to Sustainability

From an economic point of view, anything that is "sustainable" must also be "affordable," or the word would lack any practical applicability. Affordable, in turn, means that government subsidies must be discounted when evaluating the inherent sustainability of any product or process, because subsidies are both unnecessary and undesirable if something truly is sustainable. Furthermore, subsidies are subject to annual Congressional review and enactment, with all of the inherent uncertainties and delays that go with this political process. It is important to note that these key considerations are all too often overlooked or ignored by many of those promoting sustainability. Affordability is a *critical* consideration when undertaking any sustainable initiative or that initiative will necessarily fail. It is self-evident that any product, process, or service must be economical in order to be accurately termed sustainable. Nevertheless, affordability is frequently either ignored or assumed in sustainability discussions, which can result in wasting scarce resources on products or processes that lack genuine merit.

There are a number of things we can learn from classical economics that can help us to find a workable definition for *sustainability*, again bearing in mind that this word, as broadly used currently, has no consistent meaning or recognized economic basis.

- Sustainability is often used to imply that the amounts and locations of all mineral resources are known, that these resources will be completely consumed within a relatively short period of time, and that there are no available substitutes. This is an extreme viewpoint—possibly an exaggeration to "make a point"—and unrealistic.
- In contrast, classical economics teaches that each and every resource, of any type, is scarce, and only has utility and value to the extent that users/buyers can be found for it, at a price they are willing to pay.
- When demand rises, the resource supplier may produce more and/ or ask a higher price.
- If a buyer does not find sufficient value for the resource at the price offered, then that buyer will either decline to buy, use less, or make a substitution.
- If both parties find a price level at which each is satisfied, then trade continues.
- Price (or cost) is effectively a measurement of the overall energy expended both to produce and use a resource. Therefore, it is axiomatic that subsidies distort this measurement. Worse, subsidies are typically biased in favor of preferred users or suppliers over others, for ideological, political, or other noneconomic reasons.
- The world has never "run out" of any resources—typically, some resources have been priced out of some uses and replaced by another resource; for example, oil replaced coal, which had replaced wood, for use as chemical feedstocks, yet coal and wood supplies still exist and remain in use for other applications.
- Oil eventually will be priced out of energy use at some time in the future, long before it is "used up." Oil and gas as feedstocks for plastics is a high-value use, much higher than for fuel use. In the words of Sheikh Zaki Yamani, former Oil Minister of Saudi Arabia, "the Stone Age did not end for lack of stones and the Oil Age will end long before we run out of oil."

With the above basic principles in mind, think about the following:

- Subsidizing the production and use of corn-based ethanol for motor fuel attempts to hide the perverse result of consuming more overall energy than does using oil-based gasoline. Furthermore, the politically directed diversion of corn from food to fuel has caused corn-for-food prices to rise. This diversion is increasing, and the harsh impact on the poor in many third world countries has been very troubling.

- Ironically, government-subsidized diversion of corn into ethanol also harms the economics of corn-based biopolymers by raising raw material costs.

- Governments are not omniscient nor do they engage in normal commerce; typically, they have been decreeing the use or banning the use of certain materials for reasons that are based on political policy, not economic considerations. It follows that governments take no account of what might optimally replace oil in its many and varied uses. Governments are ill-advised to attempt to force economic changes and would be better off allowing normal market forces to drive the development of effective, long-term economic substitutes.

- "Alternative energy" R&D has been subsidized for many decades but has yet to produce any commercially successful sources that do not require further subsidies.

- Subsidies are justified on the basis that they "jump-start" demand, but they invariably result in long-lasting market distortions. Examples of subsidies producing a successful commercial product or process that later does not require subsidies are nonexistent.

In summary, government subsidies are a liability and not a shortcut for establishing sustainable technologies. The history of subsidies shows that while they are promoted as overcoming temporarily adverse economics of making or using something, in reality, they are all about forcing the immediate use of a costly process or product because it is politically "hot." This is usually done while hoping that research (also subsidized) may eventually find a way to overcome a currently inherent lack of affordability. Worse, subsidies typically benefit politically connected groups or individuals, who have a financial interest in conducting, expanding, and extending this same research.

A striking example of the maladroit use of subsidies and mandates is the corn-to-ethanol program for use as motor vehicle fuel. According to an article in *Time*, Feb. 14, 2011, 40% of the US corn crop was converted to ethanol in 2010, which, after blending with gasoline, became 10% of the country's motor fuel. The article goes on to state that the heavy diversion of corn from food to fuel has sharply reduced world supplies of corn for food and driven up prices 53%. Large price fluctuations are hardly unknown in commodities, but a spike this substantial and long-lasting is causing great hardship in poor countries worldwide. In 2012, the US House of Representatives began working up legislation to reduce the subsidized "food-to-fuel" use, but it remains to be seen if the Senate and the president will agree. Other examples of government green subsidies run amuck have been reported by The Center for Public Integrity and include Beacon Power, Ener 1, and Solyndra. The first two firms received US funding of $43 M and $119 M, respectively; Solyndra received $535 M. The first two filed for bankruptcy within 18 months and Solyndra within 1 month, of receiving these subsidies. It has been revealed that none of the firms ever had any customers! Needless to say, creditors of these firms will be fortunate indeed to receive even a fraction of their claims, unlike the company executives who were paid bonuses before declaring bankruptcy.

In summary, one should determine early on whether a company or green product in which you are interested is receiving a government subsidy, and if it is, you may wish to avoid or minimize further involvement because of the uncertainty attached.

6.2.2 Finding a Consistent, Practical Definition of Sustainability

The immediate challenge for management is working with individual customers who have enacted written corporate sustainability policies with which you will be required to comply. A brief survey of consumer product manufacturers' websites in 2012 revealed that there is a very broad range of positions, from virtually no sustainability policies to very detailed ones. Proctor & Gamble (P&G) is an excellent example of the latter. Most of P&G's focus is on the polymers used for disposable products and packaging; indeed, this particular end-use focus is true for the great majority of companies where sustainability is an important issue.

P&G identifies itself as the largest consumer-packaged goods company in the world today, and has drawn up long-term goals that include using 100% recycled or renewable materials for all of its products and packaging. It further defines these terms on its website as follows:

> A renewable resource is one that is produced by natural processes at a rate comparable to its rate of consumption. P&G already uses significant amounts of renewable materials in our products and packaging. However, we still use many nonrenewable materials, such as those derived from petroleum and other fossil fuels. Our vision is to use materials that are renewably sourced. Materials will come from traditional sources such as biomass and agricultural products, and research is also ongoing to understand how raw materials can be derived from biological processes such as fermentation. In addition to being renewably sourced, these materials will also be sustainable, meaning their production will not result in the destruction of critical ecosystems, loss of habitat for endangered species, or other detrimental impacts on the environment or human communities.

While this example demonstrates that P&G has a very specific vision of what they wish to accomplish under the general heading of "sustainable," it also shows that P&G has incorporated several very challenging conditions not usually found in most similar mission statements. For example, a typical definition of "renewable" materials is satisfied simply if the product is derived from plant, rather than mineral sources. However, P&G adds an important qualifying provision, namely that the material must not only be *produced* by a natural process but also the *rate of production must be equal to the rate of consumption*. This would be a very difficult criterion to ensure being met, as different, independent parties normally control the production and consumption elements. Furthermore, the term "natural process" is not defined, leading one to wonder whether this condition rules out the use of machinery for planting, nurturing, and harvesting plant materials. Thus, it is essential to discuss such criteria with customers, in order to be certain that there is mutual understanding and agreement on exactly what constitutes "sustainable" and how it is to be achieved, in order to assure compliance.

Unfortunately, there are almost as many definitions of "sustainability" as there are companies that see a need to comment on the concept. Therefore, the best policy is to be cautious when discussing this topic with actual or potential customers and ensure that both you and your customer are talking about the same thing before making any commitments.

6.2.3 Green Polymers

The central sustainability concerns for those in the plastics industry are the manufacture, processing, and use of "green" polymers, as well as their recycling or biodegradation. "Green" is a term usually applied to such polymers as polylactic acid (PLA), made from plant sources, for example, corn or sugarcane. The source of the monomer is the key to the "green" designation, so that a conventional polymer such as nylon 11, which is made from castor oil, can be considered "green," even though it has been in commercial use for over half a century without any such identification being used previously. The same can be said of nylon 6/10, as well, since the "10" monomer (sebacic acid) is also derived from castor oil.

Braskem, the Brazilian national oil company, has begun producing ethylene and propylene monomers from sugarcane. Polyethylene and polypropylene made from these monomers are therefore considered to be "green." This is a huge marketing advantage for Braskem, because it effectively allows packaging applications now in either of these two conventional polyolefins to be labeled "green" simply by switching over to these Braskem polymers. In the opinion of the author, Braskem appears to have found the most commercially viable route to supplying green polymers for most applications. Additionally, Braskem is committed to expanding its green product line beyond its current polyolefins. Unfortunately, it is not clear just how competitive the economics of sugarcane-based monomers are, compared with natural gas or petroleum-based monomers. Neither are there any published studies on the potential capacity limits of sugarcane-based monomers versus gas- or oil-based monomers. So far, Braskem is asking a premium price for its "green" PE and PP; whether these are more profitable products for Braskem than conventional polyolefins is presently unknown.

Green polymers are also associated with such other environmental initiatives as recycling, composting, and biodegradability. Virtually every thermoplastic can be recycled, as can many thermoset resins. For processors, recycling plastics materials is a practice that is almost as old as plastics processing itself. Most molders and extruders reuse their scrap or sell it to dealers who will clean and granulate it for resale, often coloring it black in order to be able to utilize varicolored resins. Such "repro" is often indistinguishable from primary resins with respect to performance. Using industrial-source recycled polymers may offer an easy, cost-effective way to be "green" for many processors, if appearance is not a major issue.

Consumer-source recycling, however, is very different from industrial-source recycling. Consumer-source scrap is a mixture of all type of waste and trash, collected by municipalities that usually only segregate scrap by general type: plastic/metal/glass, or paper. Further sorting and processing is handled by recycling companies that buy the collected waste from the municipalities. Since consumer plastics scrap is usually contaminated by dirt, food, paper, metals, etc., cleaning and sorting it can cost more than the product is worth if resold as repro.

Such scrap can be turned into useful basic chemicals or even fuel by treating it under heat and pressure to break down the polymers. Several pilot ventures are trying to demonstrate the cost-effectiveness and practicality of this approach. Compostable and biodegradable polymers are being offered for packaging applications where the materials are frequently discarded by consumers outdoors, creating litter. These two "green" terms are often thought to be more or less the same thing, but this is simply not correct.

Biodegradable is the broader term of the two, meaning the product can be decomposed by bacterial action or sunlight and moisture. Compostable means accelerated biodegradation through added moisture and warmth, usually in a confined space. Strictly speaking, recycling is a more sustainable process than either composting or biodegradation because it minimizes process losses and maximizes reusability. Composting and biodegrading turn solid polymers into moisture and carbon dioxide, which would seem to conflict with the current goal of the Environmental Protection Agency (EPA), to minimize carbon dioxide emissions. This situation could conceivably put the future of compostable and biodegradable polymers in limbo, if the EPA were to place limits on their use, although it appears that the EPA is ignoring this conflict for the present time.

Probably the industry's best known compostable biopolymer is polylactic acid (PLA), first synthesized by DuPont's Wallace Carothers (the inventor of nylon) in 1932. However, producing a high-molecular-weight version of this polymer proved to be a major challenge that was not successfully met until 60 years later, when scientists at Cargill succeeded, in the course of developing a continuous production process. Usually sold as a green alternative to conventional PET in packaging applications, PLA also has unique properties for other uses that justify its premium price more readily. One such application would be surgical implants for the human body, which break down into harmless components after the implants are no longer needed (US Patent 5,674,286). PLA is not an ideal compostable biopolymer, however, in the sense that it will not decompose readily unless processed in an industrial composting site, where higher temperatures and moisture are used than those in found in household compost heaps. In 2010, there were only an estimated 100 industrial composting sites in the United States, a very limited option. Additionally, PLA cannot be identified readily for recycling because it is a category 7 ("other") product and therefore must be collected separately from more common polymers, else it will be incinerated or buried in a landfill.

In summary, other than Braskem's line of sugarcane-based polyolefins, products made from green polymers must be considered a niche business for some time to come, assuming no government mandates are enacted for the use of such materials. Certainly there are opportunities to be developed but they should not be considered reliable long-term ones, because the basic economics and availability in commercial quantities of many green polymers are still evolving, rapidly and unpredictably.

7

Staffing for Success

It's a cliché, but it's true—people are a company's most important asset. Without the right people, the organization, physical plant, and products of a company cannot succeed by themselves. This chapter will deal with how to find, train, and retain the best people. The plastics industry has some special needs in this regard, which we shall see.

Staffing is defined herein as consisting of recruiting, training, evaluating, promoting, and firing personnel. The quality of the company's personnel must be the best that management can find or management will have self-imposed difficulties accomplishing its plans. The discussion of staffing that follows is concerned with professional personnel; plant and laboratory nonsalaried personnel are discussed later and in less detail; clerical personnel discussions are omitted as being essentially the same in any industry. Matters of compliance with government regulations are also not specific to the plastics industry and are therefore best learned from experts in this particular field.

7.1 Recruiting

How do you find new employees? There are a number of ways, all of which you will likely want to use at one time or another:

- *Classified advertisements and Internet bulletin boards*—The most commonly used. Trade publications/Internet sites are the most targeted media to use, although more general media can be useful for attracting applicants for nonsalaried positions. Never use "blind" listings. People who are already employed will not respond to them in case it is their own employer who is the advertiser. Describe your business and the requirements of the position to be filled in sufficient detail that you don't attract unqualified people.
- *Referrals from employees, suppliers, customers, friends*—Preferred because these are usually the best source for people who will fit well with the organization. These individuals will want to work with others who have their same motivations.

- *Universities, technical and vocational schools*—Another excellent source of people, but these will likely be recent graduates and therefore inexperienced—except for those graduates of "co-op" programs who have worked at plastics companies.

- *Employment and executive search agencies*—Expensive (the fee is usually 30% of the first year's salary) but sometime necessary to find just the right person to fill a particular job.

Recruiting is a tough job; the legal restrictions on prior employers' disclosure of personnel data make it very difficult to obtain a meaningful evaluation of a candidate's previous job performance. While written tests may be helpful for evaluating technical knowledge and writing skills, personality tests have questionable validity for assessing potential performance in a given position and may be subject to legal challenge. Therefore, recruiting often comes down to evaluating a candidate's experience and behavior during an interview. Since this process is too short to be more than merely an indication of whether or not an individual will become a contributing member of your management team, all new hires should be placed in probationary status for the first six to twelve months, subject to termination at will, where this is permitted by applicable state law. The objective for recruiting should always be to hire the best-qualified candidates available—usually those who have shown that they are quick to learn and willing to work hard—and never sacrificing quality for availability. It is a shameful waste of time and money to hire questionable candidates merely to fill job openings as quickly as possible. To paraphrase an old saying, "hire in haste, repent at leisure."

All offers of employment must be in writing, and the letter must state that the terms offered supersede any verbal understanding. This is not merely being professional; it is legally protecting yourself and your company from misunderstandings. At the same time, don't count any applicants who have been accepted as actual employees until the date on which they actually show up for work. It is disquieting but true that a small but significant number of people who have accepted job offers (perhaps 5%) will never actually come to work for you, and may never even notify you that they are not coming. It is also a matter of common courtesy as well as business ethics that every unsuccessful candidate who has been asked to submit a job application or who has been interviewed should be advised promptly of your decision in a kindly and appropriate way, such as a letter or phone call.

People who apply for work with your company on an unsolicited basis should receive the courtesy of a prompt reply, acknowledging the receipt of their application and notifying them whether or not you have any interest. Regrettably, a surprising number of would-be applicants, particularly those just out of college, do not take the time to evaluate whether or not a company would have use for their qualifications, for example, a distributor would not be seeking people for R&D. Don't take offense at this apparent ignorance (it's likely to be just innocence!) and trash the resume, but respond politely and point out that your company does not have such positions or it only hires people with certain experience, etc. You have a certain responsibility for building both your company's reputation and that of the industry as being a good place to work. One day that would-be researcher may find a job with another company where they will be deciding whether or not to approve your products! How you handled their job application could influence their perception of your reputation.

You should expect to see some overqualified applicants along the way. These have become an "urban legend": candidates who have more education and experience than most job openings require but are often rejected because many personnel administrators think that, if they are hired, these individuals will still be secretly looking for a better position and leave when they find one. It's been the experience of the author that such people actually do exist but they are a seldom observed species of job hunter, living mostly in "letters to the editor" columns in professional society magazines, complaining that no one will hire them. If someone genuinely overqualified turns up on your doorstep and will accept a lesser position until something better opens up in your company, by all means, give them a fair chance to prove themselves of value. Just make sure during the interviews that the reason they are out of work and can't find an opening is that their actual problem is a troublemaking personality.

The following sections will explain what to look for in a prospective job candidate.

7.1.1 Education

While the author recognizes that he is likely biased by his own educational experience, an engineering degree—or at least a degree in a hard science, for example, chemistry or physics—is an essential qualification for people seeking entry-level professional positions in the plastics industry. This is a technical business above all, and anyone who cannot quickly understand the terminology and relationships between materials and processes is likely to be lost for too long to make a successful transition into being a contributing member of the team. At middle and senior management levels, some education in business administration will be increasingly useful, regardless of the directions that one's career path takes. While a bachelor's degree is often sufficient for most positions other than research (where a PhD is usually desirable), a master's degree may be very useful in manufacturing or plant engineering. An MBA is valuable for general management as well as information management, sales, and marketing. While the author is the happy beneficiary of exposure to liberal arts courses in college in addition to an education in science, engineering, and business administration, it is usually not a good idea to recruit liberal arts majors directly from college. While a degree in, say, English literature might be useful for a few nontechnical positions, for example, advertising, one should consider only candidates who have also had progressive and substantial experience in the plastics industry for at least two years. However, experience shows that one will likely be more successful by training a "newly minted" engineer for positions in sales and marketing than by attempting to train a liberal arts recent graduate in the technical aspects of the company's products for these positions.

A word about "co-op" programs, where engineering students spend three six-month periods working in industry in between their 4 years of undergraduate study. There are a number of universities and colleges that have such programs; others offer summer intern programs that offer students shorter periods to work in industry, but these are less effective. In the author's experience, graduates from co-op programs typically exhibit more understanding of what is expected of them and have a better grasp of how to use their education in the workplace than do graduates of noncooperative programs. Should you choose to hire such graduates, then the institutions are likely to ask you to hire co-op students on a regular basis, which is usually an excellent experience for both the student and your company. Such a co-op period affords you a chance to assess the student as a possible future permanent employee, but without the obligation to hire him or her. The student also has a chance to determine whether or not this is the career opportunity that she or he wants. All in all, this is very much a win–win situation.

However, good the graduates of co-op programs are, it is most unwise to limit your recruiting to graduates of just one or two institutions. Try to blend together graduates from a number of colleges and universities, so as to avoid the possibility of cliques of alumni forming, as well as encouraging more individual and less group thinking. Even just the *appearance* of favoritism that accompanies hiring from a single institution can be harmful to morale. If favoritism does indeed exist, then this will diminish the opportunity to find and keep people motivated that have fresh points of view and different approaches to the company's business. The author has seen this happen just once, and while it was soon corrected, the situation should never have been allowed to develop and it left a bad taste for some months before it was forgotten.

Superior candidates for laboratory, secretarial, and similar positions often have two-year associate degrees from community colleges. While such an educational background may not be essential for such positions, such personnel often require less orientation, learn more quickly, and contribute more than individuals whose education ended with a high school diploma.

7.1.2 Experience

Education is, at its core, learning from other people's accumulated knowledge and their experience from applying that knowledge. However, there is no substitute for tempering and validating lessons learned in college by the reality check an individual finds through their own experience. Therefore, given a choice between apparently equally *educated* candidates, one should usually favor the one with more *experience* than the other, especially if the experience was more extensive. Most engineers start out their careers in manufacturing or R&D, which are great places to get to understand the basics of the plastics industry. At some point, however, they should also have some business experience, because this is industry, after all, not academia or government. It would not be unfair to say that the plastics industry has had more than its share of unsuccessful top managers who failed, not because they lacked solid technical knowledge, but rather that they had no significant prior experience in sales and marketing.

Take care to judge the relative quality of a candidate's experience. For example, did the individual have 5 years of varied, broadening experience, or was it actually 5 years of the same experience repeated over and over again? One way to discover the value of the candidate's experience is to ask for several examples of what that person has learned in the course of working in the industry.

Many companies recruit directly from college, with the idea that they will "fill the pipeline," replacing more experienced people as they leave or are promoted. Unfortunately, such a policy has a major problem: typically, there is a high turnover rate in this group. Perhaps as many as 30% of new graduates usually move on after their first two years of employment. There appear to be two leading reasons for this:

- College graduates who have no prior significant employment experience frequently have unrealistic expectations of what it is like to work in industry. If they cannot bring their expectations in line with the reality of the work environment they are in, they will either quit in order to find another environment more closely matching their expectations, or their work quality and quantity will decline and they will affect others' morale adversely—which will likely lead to their termination of employment. This is another good reason to hire co-op program graduates or candidates who have held several summer jobs in industry, because they will have acquired up to eighteen months of industry experience before seeking full-time positions.

- Gauging the potential performance of college graduates without a work history is difficult. Gauging the quality of the education the candidate has received is also difficult, particularly if no one doing the recruiting has actually observed recent graduates of a particular institution in action. If, despite training and counseling, a college graduate does not show above-average performance during the probationary period, he or she should be terminated. If you are willing to accept average performance, at least recruit someone with prior experience; they will likely make fewer mistakes even if their output is not the best or highest.

7.1.3 Personality Traits

With rare exceptions, it is critical to select *team* players for today's integrated management of different talents to overcome problems. The exception may be a research genius who has just the right training and experience to come up with breakthrough technology for the company. "Lone wolf" personalities, particularly in sales, manufacturing, or administration, will often cause real problems in reaching goals. Look for these traits by asking about the candidate's previous work experiences, what they liked best and least, especially about their supervisors and colleagues. People who have a history of not getting along or communicating poorly with others are unlikely to act any differently if you choose to employ them.

Occasionally, you will come into a position where you did not select your subordinates and you find that one or more of them do not meet these criteria. You certainly should attempt to work with these people to help them to modify the way they work with others, but don't expect miracles to happen; you are not much more likely to see them change than do people who enter into marriage expecting to change their spouses. After you have given them a fair chance, you will have to decide whether their particular talents are so great as to justify an exception or you should encourage them to make a career change.

Potential employees should also have a minimum level of self-confidence, based on their experience with successfully overcoming obstacles and solving problems. Otherwise, they will be afraid to take even small risks and will almost certainly make poor supervisors because they will be reluctant to delegate authority. Self-*confidence* should not be confused with self-*esteem*; the latter can arise from a false sense of accomplishment. Questioning candidates about how they have handled problem solving in the past should reveal whether or not their self-confidence is justified. Arrogance is a quality that has no value in building teamwork.

Finally, consider these thoughts from former US President Calvin Coolidge: "Nothing in this world can take the place of persistence. Talent will not; nothing is more common than unsuccessful men with talent. Genius will not; unrewarded genius is almost a proverb. Education will not; the world is full of educated derelicts. Persistence and determination alone are omnipotent. The slogan 'press on' has solved and always will solve the problems of the human race." Human behavior is one of those things that only rarely changes with the passage of time.

7.1.4 References

By all means, ask for references. Yes, they are going to be a selection of people who the applicant knows will speak favorably of him but you can still learn something. Always ask for specific examples of how the applicant carried out assigned tasks, staying away from meaningless generalizations.

For example, don't ask "How well does X get along with fellow employees?" Ask "Can you tell me specifically how X handled situations with others who disagreed with his/her ideas?" If an applicant can't furnish at least three references who are familiar with his/her work history, this by itself should be regarded as a strong caution sign. Check references by telephone; people are inclined not to answer written requests promptly, if at all. One can also gauge better if you are obtaining a truthful and complete appraisal.

Don't overlook your own contacts at companies where the applicant worked or with whom he or she came in contact in the course of work, for example, sales, purchasing, and engineering. Sometimes these sources will give you more useful information than you can obtain from anyone else. It's also a good idea to develop a relationship with one or more science and engineering professors at nearby universities, who would be willing to recommend their more promising students to you for employment.

7.1.5 Employment Agreements

Companies in the plastics industry often have need of protection from the loss of trade secrets walking out the door inside the heads of departing employees. The industry has a long, long history of employees leaving established firms to start up or join competitors. The best way to protect the company from such events is through an employment agreement. This document should include language to prevent the use or disclosure of your trade secrets to anyone not authorized by your company to have access to such information, as well as barring former employees from competing with your company for a legally reasonable period of time, for example, 12 to 18 months (longer is excessive and likely subject to legal challenge). The geographic area covered by a noncompete agreement must also be reasonable and appropriate to the employee's type of position and work assignment. For example, a salesman whose experience with the company was solely in New England cannot be reasonably kept from working for a competitor in Arizona. The scope of geographic restriction may be made somewhat broader for R&D employees, but a number of state laws do not recognize such distinctions. If a former employee can document that your noncompete agreement is preventing him or her from finding a job, despite good faith efforts to find one that is in line with his or her qualifications, then the agreement should provide for a reasonable amount of compensation for the length of the noncompete provision.

Noncompete agreements must be signed as a condition of employment *before* someone joins your firm, because there is ample judicial precedent that renders such a provision unenforceable if it is imposed on current employees as a condition of keeping their jobs. Some states, such as California, only recognize the enforceability of noncompete agreements if they are an integral part of a business acquisition contract.

With respect to your trade secrets, the best way to keep others from using them is to establish a policy that will stand up in court, showing that you have developed a process or formulation that is not widely known and that you *make a consistent effort to protect them from being disclosed to anyone not authorized to have access to them.* It may be tiresome and time-consuming, but you will not get employees to treat trade secrets seriously—or convince courts that you are taking proper steps to do so—unless you take at least most, if not all, of the following steps:

- Require escorts and visitors' badges for nonemployees entering the lab, warehouse, or manufacturing areas.
- Put up signs in the lab, warehouse, and manufacturing areas that state these locations are off-limits to unauthorized personnel (one needs an escort, pass, or key to enter).
- Remind employees from time to time, both orally and in writing, that the company does have trade secrets and employees have an obligation to protect them.
- Use coded names and/or symbols instead of conventional names and symbols for ingredients in secret formulations.
- Identify confidential documents as such by stamping them as such or using some other special identification.

With respect to this last point, do not mark everything "secret" or "confidential," or such labeling will be obvious that this is a sham. Genuinely confidential information should always be locked up in a cabinet or drawer when it is not actually in use. Customer lists in particular cannot be considered confidential unless they would be difficult to assemble and you make a genuine effort to protect them. If you do not make a credible effort to treat your trade secrets as such, a court will not do the job for you.

Technical personnel should have a clause or section in their employment contract that spells out that their assignment is to invent or create new or improved products and processes, whether they are patentable or not, and that no additional compensation is due them for executing these duties successfully. While the company is certainly not barred from rewarding employees for successful inventions and improvements, it should not be obligated to do so, either. Even so, you need to have your legal counsel check to be sure that there are no local exceptions to this rule of thumb. If you have a facility in Germany, for example, German law requires inventor employees to share in any royalties received from patented products or processes, even though they have used company time, money, and facilities to create those inventions.

7.2 Training

Hiring a new employee is a job only half done. New employees must be trained in the business and procedures of the company. Established employees also require training to ensure that they stay abreast of the technology of the industry, and that they are prepared for promotion when the time comes. Training comes about through job experience as well as formal classroom instruction.

7.2.1 Job Enrichment and Rotation

One way to add to a subordinate's experience is to broaden the assigned duties. It is often a bitter joke that such job enrichment amounts merely to more work for the same compensation—don't make this mistake. If the position carries more responsibility than before, then it should certainly carry more compensation than before. Even if the assignment is truly a lateral move, offer some benefit—a new title, a better office, etc.—to encourage the individual to make the most of the opportunity. It is a good idea to transfer promising employees between functions at intervals of about 3 to 4 years. Employees whose experience is limited to just one area are not well qualified to move into positions where they must direct subordinates in other functions. A smart, hard-working engineer who has worked successfully several years each in manufacturing, R&D, and sales/marketing has taken important steps toward becoming promising upper management material.

Job rotation also keeps people from getting stale. While individuals vary as to how long they should remain in the same general type of position before they start to lose interest and enthusiasm, a maximum of 5 years is a good rule of thumb. In small companies, where there are simply not enough positions to offer regular changes of assignment, the time frame might be lengthened beyond 5 years, but this brings into question why the company is not growing enough to be able to offer a promotion or at least the opportunity to learn something new. If the answers to these questions are not acceptable, then the company is risking the loss of a presumably good employee.

A number of companies move people around geographically in the course of job rotation, even if similar positions are available at the same site. The reasoning is usually that these firms want the employee to be exposed to different local business conditions, and that the change should be made as soon as an appropriate position is open no matter where it is. Some suspect that a few of these companies are also trying to test for willingness to accept *any* assignment given. While an occasional move is normal in the course of a business career, constant geographic relocation is emphatically *not* good practice, nor should people be penalized if they refuse such moves. This is an age of dual-career families. If one partner is offered a job in a distant location but the other is unable or unwilling to follow, then a problem has been created that will adversely affect the performance of *each* of the individuals involved. Even if the family has a single breadwinner, uprooting people can be traumatic, especially for children, particularly if it means leaving extended family members, for example, grandparents, aunts, and uncles, behind. Furthermore, management should be encouraging the development of community ties, not disrupting them. As a policy, individuals should be able to make their own career decisions, not have them forced upon them, or the company will create significant long-term problems with loyalty and job performance.

Scientists and engineers often have a general reputation of lacking sufficient people skills to be good managers. The author believes this is at best a weak generalization, but it does have at least some small basis in reality. Technical people are not inherently poor managers; it is simply that their experience in managing others in the technical sector is not necessarily easily transferable to other functions, particularly sales and marketing.

7.2.2 Continuing Education

The constant evolution of technology requires that plastics engineers and polymer scientists must continue to stay abreast of these developments by reading technical journals, attending conferences, and taking courses. Establish a company library, buy reference books for it, and subscribe to industry journals that your employees can access and read. Depending on the preferences of the employees who will use them, these books and journals can be electronic or hard copy; both might be the best solution. Encourage your employees to attend and participate in professional societies and technical conferences as well as take continuing education courses at local colleges and universities. The company should reimburse at least a portion of the tuition charges as long as the employee earns a passing grade. While it is always a good idea to fill in any subject areas not taken during one's previous education, the main objective should be to keep up with the latest changes in technology. There is also much to be said for taking a program that leads to an advanced degree. This ensures that the course of study is comprehensive and enhances the performance potential of the individual. However, a common error is not to recognize the individual's new status when he or she attains their degree. If they are not promoted, given a new assignment, or at least given an appropriate pay increase, they may believe that their performance is being taken for granted and lose ambition or even eventually leave. Of course, there is always some risk attached to training and educating personnel for higher-level positions; they may find the grass greener elsewhere, or they may turn out to be unsuited for higher-level management. Nothing in life is risk-free, but not training people to become better employees increases the risk that they will not perform their duties as competently as they might, to the disadvantage of the company.

The company should also conduct in-house seminars and courses. For example, this can be a very cost-effective way to improve planning methodology, using a common approach throughout the company. Also, when new information technology is adopted, it is essential to train everyone on how to utilize it. It is also a good idea to encourage your technical personnel to become members of technical societies, attend conferences, and, where appropriate, present papers. Such surroundings offer opportunities to learn from peers and experts on the latest advances in plastics technology. Serving on boards and committees of technical societies is a great way to hone people skills, too, and is part of being a professional engineer or scientist. If you can afford it, inviting experts to conduct a seminar or give a presentation at your company is another excellent way to keep your technical staff up to speed on the latest developments.

Some managers complain that if they pay to train or educate people, then another company will hire them away. Think about that for a moment—do you really want people that *no one else wants*? Yes, there is always a risk that you will train someone, and they will repay that investment by leaving. That risk must be balanced against the superior performance you will get out of trained people who do stay, or even out of the people who stay only for a while before they leave. This shouldn't require much thought; training pays for itself many times over, even allowing for some attrition. If your company's turnover is excessive, it isn't training that's causing it! The last section of this chapter deals with this latter problem.

7.3 Compensation and Reviews

While there may still be some firms in the industry that compensate their managerial and professional employees solely through straight salary, none come to mind. Almost every professional above entry level today is compensated through a combination of salary and incentive. A combination of performance-based cash bonuses (for short-term goals), and stock options or stock appreciation rights (for long-term goals) is common.

Compensation plans need to be based on *national* surveys because you are competing for people throughout the country. Regional cost-of-living differences sometimes may require adjustments, but these should be handled outside of the basic salary and incentive compensation system.

Salary reviews should be conducted annually and without fail. Companies that permit reviews to "slip" beyond the scheduled date are not treating their employees fairly and often lead to diminished performance and turnover. If the company is in serious financial trouble, then it may make sense to announce that there will be a freeze on compensation or even a salary cut, but such drastic measures must be only for a specified period of time, and certainly not exceeding one year.

Salaries need to be fitted into an ascending scale of position responsibilities. The salary structure should be updated at least annually to ensure that it is at least comparable to the company's main competitors. At the top end of the scale, make provision for your best-performing professionals to win additional compensation without the necessity of becoming managers. Some of your best R&D scientists and salespeople, for example, may only be marginal managers. Don't penalize them financially for staying in positions where they contribute more than they could as managers.

Based on the author's experience, a compensation committee is recommended, even though it is not a standard practice in many companies. The author has found that having more middle management participation in setting compensation standards tends to reduce the "us versus them" mentality often found separating upper management and subordinates on compensation issues. It also helps make the process of allocating a fixed amount of money less arcane and more understandable to participating department heads, who then can represent the compensation process more accurately and fairly to their subordinates than otherwise might be the case.

R&D personnel with patentable ideas present a special concern: are they paid to produce such ideas or should they receive special financial incentives for successfully patented ones? As discussed elsewhere, not every good idea needs to be or should be or even *can* be patented. In the experience of the author, giving recognition to the inventor of a successfully patented product or process should be primarily social, such as holding a dinner and giving the inventor a framed copy of the patent, or nominating the inventor for a scientific society prize. The value of successfully applied new ideas, patented or not, can be recognized by bonuses, salary increases, and promotions, in this order and as appropriate in relation to the value created by the invention and the particular talents of the individual involved.

7.4 Promotions

Next to making more money, almost every employee wants to be rewarded for good performance with a promotion. However, while promotion is not always an appropriate "reward" for strong performance, money is! Not every high-performance employee is necessarily promotable. The famous "Peter Principle" is ignored at great risk: it says that people tend to be promoted to the level of their incompetence. The cardinal principle that must be at the forefront of your thinking is that you must promote based on your assessment of the individual's *potential capability to do the new job, and not as a reward for doing their current job well.* This may well be a difficult, but not an impossible, decision, but it is essential that you make it, based on what is best for the company and the employee.

When people are promoted to their "level of incompetence," you have a lose–lose situation. You do not get the performance you expected and either they will become miserable and quit, or you will wind up firing them. Thus, you will lose a previously superior performer. Only in large companies and with very resilient people is it generally possible to transfer or demote someone who does not work out, back to their old level and retain them.

Some firms have a "fast track" system for identifying and promoting people who have been identified early in their careers as having high potential. The idea is to move them up the ladder as quickly as possible. This is a faulty idea with high-risk consequences. First, people progress at different rates during their careers; there are very, very few individuals who can master every position in which they are put, in a short period of time. Second, proper management development must include the idea of *seasoning*: experiencing the ups and downs in a position that usually do not conform to any rigid time frame. Third, most people are not stupid; they can usually spot when someone is getting preferred treatment. This shortchanges both those who are on the fast track and those who are not, because the suspicion has been planted that the fast trackers have not earned their promotions on the same basis as everyone else. This is destructive of good working relationships all around. The author has never seen "fast tracking" genuinely benefit a company overall but has seen it damage those that used it.

7.5 Firing and Laying Off Personnel

Employees are generally fired for one of two reasons: for "cause"—they have broken government laws or violated company rules—or for "unsatisfactory performance." The first reason is atypical, and most managers will not shrink from handling the situation properly. If an employee requires termination because they have done something seriously wrong, such as stealing, fighting, damaging property, intentionally or carelessly violating employment or environmental laws, etc., they should be suspended from work immediately while you verify the violation. Once the violation is confirmed, then they should then be discharged at once; only allowing them on the premises to remove their personal possessions in the presence of security guards. A violation of government law should be reported to the proper authorities for their action; failure to do so could conceivably expose the company to obstruction of justice charges.

Unsatisfactory performance also should be treated in a straightforward manner, although the urgency and concern for security may not necessarily be as compelling. If an employee receives a second (or third) unsatisfactory performance review, each of which specified what the person under evaluation had to do to bring themselves up to acceptable performance and the consequence of failing to do so, then it is time to "require a career change." Firing someone for other than "cause" is usually a very distasteful job for most managers because it is only human to feel sorry for the individual being terminated. Nevertheless, it is an extremely important job, for several reasons:

- If your performance review system is to have any meaning and integrity at all, then satisfactory reviews must result in retention, and successive unsatisfactory reviews must result in dismissal. This is fundamental to personnel management and to the credibility of your administrative system.

- People who cannot or will not contribute at an acceptable level of performance are never happy in their job. This alienation infects others and pulls down *their* performance as well. It is vital to maintaining the morale of contributing employees that you dismiss chronically underperforming and disaffected employees.

- It is ethically wrong to keep someone in a job that they cannot or will not execute on a satisfactory basis. It is a rare situation where such individuals can be transferred within the company to another position where they can perform, because there is usually too much baggage carried along to make this feasible. Assuming the situation is not one of those rare ones, then you must fire that person, telling them exactly why, and with a reasonable financial severance package. You may wish to arrange for an outplacement service, depending on how likely the individual is to find another position within a short period of time. While it may sound facile to state that a manager is actually doing a nonperforming employee a favor by forcing the career change, it is nevertheless true. A surprising number of such employees will actually express a sense of relief when they are required to leave their jobs. People often need the push to move on to another job because they resist change; sometimes they are "in denial" of their failure to perform. Handled properly, for example, with outplacement counseling, the fired employee gets a fresh chance to start over again and regain the self-confidence and self-esteem that they had lost in their previous position.

A particularly repellent reason for dismissal of personnel was instituted several decades ago during the brief tenure of one of the automotive "Big Three" CEOs, described as similar to one used by Jack Welch at GE. It consisted of mandatory classification of salaried employees into A, B, and C categories, in predetermined percentages, in the course of their regular performance evaluations. C category personnel would be denied bonuses, and if they remained a C after a second review, terminated. Never mind whether one or more of these groups of employees had been previously selected with care to include only top-level performers—off with their heads! This procedure earned the company a series of high-profile individual and class action age, gender, and race discrimination lawsuits, not to mention a major hit to morale. The company eventually settled these suits rather than let them drag out through the courts with attendant bad publicity. As described elsewhere, no lasting good comes of setting employees against each other; the team cooperation required for successful business operation will evaporate in a cloud of distrust and backstabbing. Furthermore, any system that deliberately stigmatizes an arbitrary percentage of the workforce is contemptible and has no place in any ethical company.

Layoffs due to restructuring are a very different matter than discharging unsatisfactory performers. As a matter of principle, layoffs should not be used to deal with a temporary contraction in the business cycle. If the company is in trouble because of overstaffing, then among the first to go should be the executives whose poor judgment created the situation. Every company should be run on a sufficiently lean basis that there is no "fat" that can be cut in down cycles. Expansions should be based on improving productivity or even temporarily outsourcing, wherever possible, and not just mindlessly adding expendable bodies. Layoffs are a desperate measure and should be used only as a last-ditch, emergency solution to save the company and the remaining jobs. Morale will suffer from a layoff, and the reasons must be clearly and convincingly communicated to the surviving staff or they will remain demoralized for a longer period than the company can really afford. Layoffs are not cheap, either; severance costs can easily exceed short-term cost savings. If layoffs cannot be avoided, then management *must* ensure that key personnel are not lost in the cutback. Across-the-board layoffs are a shabby way to avoid dealing with the problem of who is to go; do not do it. Any layoffs must have a rationale that everyone can understand, so that employees can understand that anyone laid off will be recalled as soon as business conditions permit.

Some companies have tried a "share the pain" approach, by applying compensation cuts to everyone rather than laying people off. This may make sense if personnel compensation has been reasonably generous in the past and the cuts are not drastic. It may also have merit if the company's staff is small and it is impractical to reduce the number of employees without a serious adverse impact on operations. The technique is more likely to work if the time interval that has to be bridged is less than a year, recovery seems highly probable, and the cuts are restored at the earliest opportunity. Employees who have been through such situations often have stronger positive feelings about the company and each other than in companies where layoffs were carried out. Of course, it is essential that *everyone* share in the cuts, especially the managers. There is a significant risk to this approach; however, because your better performers are apt to leave for higher-paying jobs if the cuts last for more than 6 months, unless the general economy is in bad shape for an extended period, such as during the Great Recession that began in 2008.

An alternative approach in this same vein of "share the pain" is to ask for volunteers to be laid off. The layoff period should be short, usually not more than 90 days. Volunteers should be guaranteed that not only will they be brought back, but that they will receive a significantly better severance package in the event that permanent layoffs prove necessary later, say, within 1 year.

In past economic declines, a number of companies in the industry have managed "downsizing," for example, layoffs, by offering enhanced early retirement packages to their older personnel, generally those who are 55 years of age or more. Since anyone over the age of 45 is protected by federal employment antidiscrimination laws, this procedure cannot be easily limited to weeding out the less productive employees—everyone in the same category has to have the same opportunity offered to them. As a result, a number of more desirable employees have been lost along with the less desirable ones. In the opinion of a number of outside observers, the mass early retirement of a "generation" of experienced, highly competent senior personnel was a particularly bad bargain. The companies may have been able to reduce its annual compensation costs in the short-term, but they paid a heavy price—twice—for this action that more than cancelled out any savings overall. The most obvious price was the cost of the enhanced retirement, which frequently required the companies to take write-downs against earnings, typically more than the amount of earnings for one or even two quarters. The less obvious cost came about from operating problems that were likely to have been avoided if the more senior personnel had been on hand to perform or advise. A few companies realized this and actually rehired some of the retirees as consultants—which might have been a better solution than layoffs initially. Overall, it would seem that the "early retirement" approach, as practiced some time ago, was too drastic and inflexible to be useful.

If you are looking for part-time workers, however, you may find that some of your retirees would be interested as long as their pension benefits were not adversely affected. This is a great way to utilize people that you know and who have valuable experience to offer (see more on this in Section 7.7).

Finally, perhaps the most important reason for not using dismissals to deal with temporary contractions in the business cycle is that you will *need* those trained and experienced personnel to handle business when business conditions improve again. Remember how expensive and time-consuming it was to find, hire, and train new people? If you cut staff in a downturn, you will be faced with having to go through the same cycle of recruiting, hiring, and training all over again—only this time, it will be more difficult to get top people, because you now have acquired the undesirable reputation of "quick to hire and quick to fire."

7.5.1 Firing and Laying Off Outside the United States

It is truly exceptional for companies to freely terminate employment or lay off personnel in countries other than the United States. In Europe, Latin

America, Africa, the Middle East, and much of the Far East, it is generally both difficult and expensive to do so (not to mention time-consuming). Particularly, if you are starting up an operation in Europe, you may wish to consider contracting with a firm that specializes in furnishing temporary employees to companies (this is discussed from a US point of view in the next section). Often, such companies will let you convert one or more temporary employees to permanent ones, if need be, for a fee that is comparable to using an employment agency. Germany also has regulations that permit hiring people as trainees for up to 2 years, before one has to make a decision to offer such employees permanent jobs.

7.6 Using Temporary Personnel

By no means does every position in a company need to be filled by a full-time employee. There are many jobs that can be performed by a part-time employee, temporary or contract worker, or consultant. Project work, such as market research, customer satisfaction surveys, and even some technical service, for example, fielding telephone inquiries, can be successfully accomplished via this route. Also, it is often wise to use disinterested, outside parties to confirm critical data that has been generated internally, such as project work. Do not use such people in jobs where you need to build up a permanent reservoir of experience within the company.

Some company activities can be compartmented and contracted out, as one way to deal with the up cycles, yet not creating an oversupply of people to deal with when the business cycle turns down. Some aspects of manufacturing can be handled on a contract basis, for example, polymer producers can utilize contract compounders, certain types of plant maintenance can be done by independent contractors, use contract trucking instead of or in addition to your own fleet, farm out molding work or decorating to other molders, use an outside payroll service, etc.

The primary advantage of using temporary or part-time personnel is that they can be laid off instead of your permanent, full-time employees, without the morale, financial, and reputation problems cited earlier. Do not be fooled by the stereotype that temporary personnel are necessarily less expensive than permanent ones; competent professionals command fairly high levels of compensation. Do not begrudge this because they are necessarily earmarking some of it for tiding them over between job assignments as well as providing for their retirement and health care benefits that you are not paying for directly.

Retired personnel are sometimes interested in part-time or contract work. It makes sense to utilize retirees wherever feasible to do so, as their loyalty, experience, and past performance are known qualities. It is also a good morale booster, because it shows that management appreciates these people for their ability to continue to contribute, albeit on a reduced basis.

While consultants have acquired an uncertain reputation in some quarters, this is an undeserved generalization. Of course, there are good and bad consultants just as in any field, but the less competent ones are quickly weeded out after a few years; they simply never get repeat work, which is critical to succeeding as a consultant. There are many instances where hiring a qualified consultant for a specific assignment is a much sounder move than handling the project internally. Consultants often can bring more expertise than exists in-house to bear on specific problems, which is particularly helpful with nonrecurring issues. It has been the author's experience, both as an executive and as a consultant, that individual consultants can be found that have more specialized expertise applicable to the problem at hand, and cost less, than can be found among the employees in large firms. Consultants should not be brought in to provide justification for a decision that management has already made; this is a misuse. Sometimes, members of the board of directors can contribute from their own knowledge and experience, as unpaid consultants—unpaid, in the sense that they would not receive additional compensation beyond that which they receive for serving on the board.

In Europe and other countries with laws that make it very difficult to dismiss employees, it is commonplace to use contract workers. The exact length of time varies from country to country, but you cannot keep the same temporary worker on for more than perhaps a year or so, because they will—by law—then become permanent employees, so this requires regular follow-up to avoid being surprised.

7.7 Retention

Finally, a word about retention: keeping people. Considering how much time, money, and trouble it is to find, hire, and train good, qualified people, it is amazing how few managers think regularly about how to *keep* those people. While it is true that no one is indispensable, it is foolish to ignore the need to keep people when it is really not all that difficult to do so. Corporate loyalty in the days of downsizing and restructuring is not what it once was, but management can and should find ways to repair that damage.

Why do people stay? The reasons are sprinkled through the preceding sections of this chapter, so the following list pulls them together, in order of importance:

- People enjoy career growth and see it continuing.
- People like their jobs; they find the work interesting and meaningful.
- People like their fellow employees; there is a team spirit.

- People think that management treats them fairly, and they have an opportunity to express their views and influence decisions.
- People think their pay and benefits are fair. (Believe it or not, study after study shows that this is *not* the first consideration, or the second or even the third—it comes *after* all the others. Also, notice the word "fair"—not "higher than anywhere else." Now, that is *not* an invitation to underpay people, but it certainly does show that people are not generally greedy, but are more interested in being treated *equitably* than in how much they make.)

Recognition of a job well done should never consist entirely of financial rewards, which are, of course, very important, but need to be treated privately. There are many ways to make public that someone has done a particularly effective job: some extra time off, an award plaque, special mention in front of colleagues during an appropriate meeting, just to name a few.

7.8 Plant and Laboratory Nonprofessional Personnel

To the casual eye, nonprofessional personnel, for example, lacking a bachelor's degree or higher in "hard" science or engineering, for manufacturing and the laboratory are more or less interchangeable, as long as they are given some basic training. This is a mistaken concept. The potential for expensive spoilage or waste, even for hazards to life and the environment are significantly greater in our industry than in many others. It is essential that those entrusted with the operation of equipment and disposal of materials are highly responsible and trustworthy.

Plant and lab people in the chemical and plastics industries must be hired after a careful screening to ensure that only those who are particularly conscientious in following established operating procedures are hired. Furthermore, these attitudes must be fostered by training and supervisory reinforcement. Willful disobedience of these procedures or a negligent attitude cannot be tolerated and appropriate disciplinary procedures followed, including (if warranted) dismissal.

Since plant and lab employees are expected to be a cut above other non-professionals, they should be paid more than an average wage for parallel jobs in the area. Nonprofessional personnel tend to be less mobile than professionals are, so that compensation needs to be compared against the area or regional levels in the industry (effectively there are no national levels per se). You do not want to lose people that have taken much time and money to find, hire, and train.

Especially for lab positions, it pays to look for people with at least some education beyond high school. For example, an associate's degree in chemical technology or a bachelor's degree in biology can be a good indication that an individual has the interests and training to become proficient in your laboratory, even if these subjects do not bear directly on plastics. An associate's degree in plastics technology would be even better, of course, but there are only a limited number of institutions conferring such degrees.

7.8.1 Unions

What if a union tries to organize your nonprofessionals? The first question to ask yourself is, why? Unions usually enter the picture as the result of poor employer–employee relations. Since you are presumably paying more than the average wage—and while unions often promise higher wages, money is seldom the main reason for unrest. You should take a hard look at how your supervisors and managers are treating hourly employees to see if this is causing problems. If there is, fix it promptly! By the way, this kind of potential problem should always be on your radar, long before a union representative shows up. Once a union finds enough support to call for an election, any steps you take may already be too late.

While it is possible to have a good, constructive relationship with a union, they do add cost to operations through inefficiencies (work rules), complaints ("grievances"), and even strikes. It is quite common for one or more disgruntled employees to bring in union organizers, participate prominently in the campaign for recognition, and then leave within a few months after the union has been established.

Unions found their initial justification in the harsh exploitation of factory labor, following the Industrial Revolution in the 18th century. Unions came into being in the 19th century under the leadership of socialist reformers. Aided by government in the early 20th century, unions were able to negotiate for improved working conditions, and higher wages. In the latter half of the 20th century, governments have taken over the regulation of working conditions, as well as such benefits as pensions and health care. Additionally, the post-World War II industrial boom increased demand for workers, with an accompanying rise in wages and benefits. Over the years, most national industrial unions have found it increasingly difficult to justify the substantial dues they charge and have seen their membership ranks decline sharply. As a result, they are eager to find new opportunities to organize even small businesses. Independent unions (whose membership is limited to a single company's nonexecutive personnel) are, in the experience of the author, far less contentious in negotiations, charge lower dues to their membership, and achieve virtually identical results versus national unions.

Once a union has been certified as the bargaining agent for your employees by the National Labor Relations Board, it is extraordinarily difficult for it to be decertified—and company management is barred by law from having any role whatsoever in decertification. Unions have fallen into disfavor among manufacturing employees in recent decades—US Department of Labor statistics for 2011 indicates that fewer than 7% of US privately employed industrial workers are unionized. These are mainly legacies from the past, and limited to large, old companies, such as automakers and steel mills. Unions presently win considerably less than half of the organizing elections that take place, but it is an expensive, time-consuming, and contentious process, regardless of the outcome. This is one case where the proverbial ounce of prevention is very definitely worth a pound of cure.

8

Tools for Management

8.1 Analyzing Your Business

The first step toward taking control over the direction and success of your business is to understand what actually *constitutes* your business, by analyzing its components and how they contribute to or detract from the whole. These are both qualitative and quantitative analyses. Only then can one decide intelligently which areas to emphasize and which to de-emphasize or even discontinue. Making such decisions imposes opportunity costs, of course, so that they must be informed and considered judgments, not intuitive ones; sometimes the correct decisions prove to be counterintuitive. The term *opportunity cost* is used to mean that pursuing one option necessarily means giving up the opportunity to pursue another, since no one has unlimited resources. As a consequence of deciding what your best opportunities are, you will need to assign your best people resources to these projects. The process described in the following paragraphs describes a basic set of tests concerning how your company's resources are being applied and where you may benefit from redirecting these resources. Only after you have created such an appraisal (and the competitive analysis described later in this chapter) can you undertake to create business plans.

One school of thought suggests that if you are not number one or two in a particular line of business (in terms of sales) and have little likelihood of reaching either of these positions, then you should exit the business. Implicit in this philosophy is the idea that if your market position is large enough, high profitability will follow. While there is some truth to this, it is an oversimplification of the relationship between sales to profitability. Nevertheless, it is important to know if your market share by product (or service) is increasing, holding steady, or declining. The first instance is good, the second may be acceptable, but the third is a danger signal and requires action.

Another area of critical concern is new product introductions. Too often, the seductive excitement of bringing out a series of new products leads to glossing over a key goal: will they make a greater return on investment than you made before the introductions? *The single most important thing you can do when introducing a new product is to establish positive cash flow by not later than the end of the first year.* One might make a special case for an exception to the 1-year rule, but anything over a year is on shaky ground because the reliability of the forecast is significantly less. This is a more important principle than simply establishing sales volume. If a new product proves to be unprofitable despite the prior analyses, you need to detect this flaw early on before you have committed large sums of cash (and time) that are likely to be unrecoverable. Either fix it immediately or drop the product and move onto the next most promising one.

8.1.1 Current Relative Profitability

The quantitative business analysis begins with determining the relative sales revenues and corresponding profitabilities of the products and/or services that comprise your business. Professor Peter Drucker in his book, *Managing for Results,* recommends drawing up a table, in which the company's products (or services—for purposes of this chapter, we will assume that the two are interchangeable)—are as follows:

- Individual product sales revenues, net of the cost of raw materials.
- Each product's percentage of the company's total net sales revenues.
- Each product's cost burden (more on how to determine this in a moment).
- Percentage of this individual product's burden of the company's total costs.
- Each product's net earnings (net sales revenues—per the first item above—minus manufacturing and overhead costs).
- The percentage of the product earnings of the total company earnings.
- The product's "contribution coefficient" (a value obtained by dividing the percentage of its net earning by its sales revenues).

Professor Drucker defines revenues and costs a little differently than many accountants would for this exercise. First, he uses a "value-added" approach, meaning sales revenues are stated net of raw materials costs. Second, he allocates manufacturing cost and general overhead (sale, marketing, R&D—whatever is needed to develop, make, sell, and maintain a product) on the basis of *time* or *transactions,* not just physical volume (tonnage). What might these factors be in the plastics industry? Here are a few examples:

- For a polymer producer, compounder or processor, the manufacturing transaction cost would include the personnel assigned to operate, and maintain the equipment necessary to produce a single order (including a production lot made either to fill accumulated orders or to build inventory).

- For a distributor, the service transaction cost would be the time needed by customer service to process an order, either from stock or by ordering from the manufacturer.
- For an equipment manufacturer, sales transaction costs would include the time spent preparing the average number of proposals required to obtain a single order.

This definition of costs is likely to require some estimating on your behalf to get the most representative numbers, but ultimately it will be more significant than conventional cost allocations based solely on volume. For a polymer producer or compounder, the costs associated with processing an order for 1 ton of material are not much different from processing an order of 20 tons. Drucker considers the cost of each of these orders as equal, but conventional cost accounting would make it appear that handling the larger physical volume order only costs a fraction of what it costs to handle the smaller order. When very small orders are involved, setup costs may become equal to or greater than running costs. Each method makes compromises, but Drucker's approach requires that you examine costs from a fresh perspective and learn something in the process.

Table 8.1 illustrates this concept, as a series of hypothetical products and end-use categories, found at an injection molder. The column headings are self-explanatory, and it is instructive to see the profitabilities relative to the cost burdens and sales revenues. Note the "contribution coefficient" in the last column, which measures the ability of a product to generate income as its sales volume displaces that of another product or products. In our example, the product category with the highest contribution coefficient is industrial parts. Assuming that there is equal demand for each product, then every dollar of cost and resources invested in increasing the sale of industrial parts will return more than twice as much profit than a dollar

TABLE 8.1

Current Profitability Analysis

Product	Sales, M $	% of Total	Costs, M $	% of Total	Earnings, M $	% of Total	Contribution Coefficient
Automotive	40	35.4	38	37.0	3	26	0.7
Housewares	35	31.0	28	27.3	7	61	1.7
Electrical/Electronics	15	13.3	11	10.7	4	35	2.3
Industrial	8	7.1	5	4.9	3	26	3.3
Medical	7	6.2	6.5	6.4	0.5	4	0.6
Specialty packaging	5	4.4	6	5.9	−1	−9	−1.8
Recreation	1.5	1.3	3	2.9	−1.5	−13	−8.7
Construction	1.5	1.3	5	4.9	−3.5	−30	−20.0
Total	113	100.0	102.5	100.0	11.5	100	

invested in boosting any other product line. The lesson here is self-evident: put more of your resources to work increasing sales of industrial parts, and your earnings will rise sharply. However, this is not the whole story. Some of the products losing money at present are worth supporting as we will see a bit later.

8.1.2 Relative Profitability Potential

Next is a qualitative analysis of the product lines, to categorize how they differ from each other with respect to future profitability. In effect, we are estimating where each product line would be located on a classic S-shaped curve depicting the logarithmic acceleration, then deceleration of sales from the time a product is first introduced to the time it is fully mature. We can hardly improve on Professor Drucker's terminology, so we have assigned products into the following categories:

- *Today's breadwinner.* A product in this category shows significant volume, but its growth is starting to slow—the second inflection point on the S curve. Its contribution coefficient is average. Obviously you need to maintain this product; it is paying a lot of your bills, but it is now at the stage to be "milked," that is, allocate sufficient resources to maintain a stable market share and profitability, but nothing more.

- *Tomorrow's breadwinner.* This type of product is already important but is just beginning to experience its main growth. Its contribution coefficient is normally high, and it is at the first inflection point on the S curve. Assuming it has become profitable, its contribution coefficient will be high. Putting more resources into building, this line should pay off increasingly with time.

- *Yesterday's breadwinner.* This product is marked by high sales volume but softening profitability and low growth. Its contribution coefficient is low, as it is at the end of the S curve. Often "the product that built this company," it now requires price cuts to maintain market share or even keep it alive. This product should be not only be milked, but also considered for sale or even termination.

- *Repair job.* A product that exhibits substantial volume and good growth potential (near the middle of the S curve), but has low profitability that stems from a single major defect, a problem that is clearly definable and readily corrected, such as positioning in another market. Fix it, and "let the good times roll!"

- *Productive specialty.* A product that has a distinct but limited market, with high profitability and needing only limited resources. These are keepers.

- *Unnecessary or unjustified specialties.* Products with more variations than are really needed by the customer or that are sufficiently undifferentiated from others in that they cannot command a price premium. Try consolidating or selling off these products; terminate them as a last resort, and redeploy the resources previously used to sustain them.

- *Development products.* Products still in the introductory stage and, though as yet unproven, appear to have potential. The technology is cutting edge, but the economics are as yet unproven. Often such products do not receive adequate support because to do so would mean redeploying resources from or more of management's favorites: yesterday's breadwinner or an investment in management ego. Nevertheless, be careful about committing so much support that these products will be unable to achieve positive and improving cash flow after 1 year.

- *Failure.* Self-diagnosing—like the infamous Ford Edsel, rejected by the market almost as soon as it is introduced. Ford had the good sense to see this and drop the product instead of converting it into an investment in management ego. Terminate as soon as possible.

- *Investment in managerial ego.* A product that should be a success—but is not. However, management is so convinced that it "is the best in its class" that it keeps pumping resources into the product. The poorer it does, the more management gives to it, a sort of death spiral that, in extreme cases, can actually suck the life out of an entire company. Unfortunately, investments in management ego are not a rare phenomenon. Someone needs to tell management the cold facts. Make a quick fix or drop without further delay!

- *Cinderella* (or, a "Sleeper"). A product that might do well if it had adequate support. Often such products do not receive support because to do so would require drawing down support from one or more of management's favorites: today's or yesterday's breadwinners, or an investment in management ego. The solution should be obvious.

This method of categorizing products is illustrated in Table 8.2, following from the examples shown in the previous table.

8.2 Benchmarks for Allocation of Costs

A good way to judge how you are allocating costs is to rate your company against industry averages. Knowing how your competitors allocate resources

TABLE 8.2

Category Checklist

Category	Sales Volume	Sales Growth	Contribution Coefficient	Action
Today's Breadwinner	High	Slowing	Medium	Monitor
Tomorrow's Breadwinner	Medium	Growing	High	Support
Yesterday's Breadwinner	High	None	Low	Milk or sell
Repair Jobs	Low	Low	Low	Fix or drop
Productive Specialities	Low	Varies	High	Milk
Unnecessary Specialities	Low	None	Low	Drop
Developmental Products	Low	Unknown	Unknown	Support
Failures	Low	None	Low	Drop
Investments in Management Ego	Low	None	Low	Drop

would be even better, but this information is sufficiently sensitive that it is not often public knowledge. Averages can be misleading because actual numbers vary according to the type of company, its size, type of financing, growth rate, products, and competitive pressures, so it is necessary to recognize that they are only indicators. Nevertheless, let us look at a few examples of what might constitute an average or typical allocation of costs. While return on investment is more important to business owners than return on sales, this type of information is not easily found, varies widely, and is therefore omitted.

8.2.1 Polymer Manufacturer

Stand-alone numbers for polymer manufacturers essentially are unavailable, since almost all of these operations are integrated parts of large chemical or petrochemical companies. If data for a firm existed, then a "typical" integrated polymer manufacturer producing a semi-commodity product during an expansion cycle of the economy might find that its expense breakdown looked something like this, with costs expressed as a percentage of sales:

Raw materials	34%
Manufacturing	35
Gross margin	31
Administration	3
R&D	6
Sales and marketing	4
Net income before interest and taxes	18%

The relatively low manufacturing cost shown here is typical when making long production runs of a limited number of products. It also includes depreciation, which may not accurately reflect the replacement cost of the facilities, depending on their age, inflation rates, and how current the plant's level of technology is. This is a critical detail, because management may not realize this and seek to raise market share by cutting prices, but then learn later that insufficient cash flow is being generated to finance growth, plant upgrades, or even adequate maintenance. Many polymer producers target a return on an investment of 35% or more to justify building new plants, although it is questionable that they actually do so, on average. R&D costs are higher than in other plastics industry sectors because polymer product and process development require scale-up steps that are not found in compounding, molding, etc., and often require pilot plant facilities.

8.2.2 Compounder

Compounders usually work on smaller gross margins than polymer manufacturers and therefore must have lower overhead expenses. A specialty compounder might have an expense breakdown such as this:

Raw materials	58%
Manufacturing	18
Gross margin	24
Administration	3
R&D	2
Sales and marketing	5
Net before interest and taxes	14%

These data are also somewhat hypothetical, as there are no genuinely "pure-play" publicly held specialty compounding companies. There may be some blurring between marketing and R&D costs in the area of application development, as R&D largely consists of fine-tuning products for custom applications.

No meaningful current information is publicly available on the return on an investment in compounding, but values of as much as 35% have been realized during good economic periods in the past decade or so. However, this number is influenced strongly by the age, condition, and type of the compounder's equipment. If the equipment is relatively old or purchased used, and single-screw, then returns will be on the high side. Conversely, if the equipment is new and twin-screw, returns will be somewhat less, unless the more sophisticated equipment is used only for high-end compounds, for example, PEEK-based.

8.2.3 Distributor

Distributors have a wide range of gross margins depending on how they are compensated (resale versus commission), the types of materials they handle, the industries they serve, and their size. In the example below, it is assumed that a small distributor buys engineering plastic resins from several manufacturers, stocks them for a predictable, nonautomotive clientele, and resells.

Raw materials	75%
Storage and shipping	6
Gross margin	19
Administration	3
Technical service	1
Sales and marketing	6
Net before interest and taxes	10%

Distributors specializing in small lots of commodity polymers ("hand-offs" from suppliers) may show lower raw material costs and higher margins. With favorable credit lines from suppliers and prompt payment by customers, some distributors should be able to turn over their inventories within 30 to 60 days, minimizing working capital requirements. If they do not own warehouses, silos, blending and packaging lines, trucks, etc., they can show relatively high returns on invested capital. However, high returns are only justified by high risks, and if the inventory does not move promptly and at the expected price, the distributor will be squeezed. While polymer producers sometimes provide technical support with respect to complaints about the performance of their materials, the distributor almost always has to provide "first aid" when it comes to processing problems. This can often be addressed by using a consultant or having one of the lab technicians or sales representatives handle troubleshooting as a secondary responsibility.

8.2.4 Processor

Processors' expense allocations can vary even more widely than those of distributors, depending on their size, the extent to which their operations are fully integrated, and whether they do custom, proprietary, captive work, or some combination of the foregoing. The following example is for a custom injection molder and assumes smaller size and limited integration, for example, mold maintenance, but not mold building, assembly but not decorating, etc.

Raw materials	35%
Production	45
Gross margin	20
Administration	3
Technical	3
Sales and marketing	4
Net before interest and taxes	10%

These are purely "ballpark" numbers because the size and types of processors vary so greatly. It would not be unreasonable to estimate that a molder with a sound business model could realize a net return on invested capital of 20% to 25%.

8.2.5 Machinery Manufacturer

Machinery manufacturers are subject to ups and downs of demand that match the business cycle, only more exaggerated. In the past decade, most of these companies have been merged into megacompanies, and there is essentially no current financial information available that breaks out plastics machinery. Ten years ago, when there were still a few independent injection molding machine makers, a snapshot of the financial picture of such a company during a good period might have looked like this.

Production (including raw materials)	75%
Gross margin	25
Admin, R&D, sales	15%
Net before interest and taxes	10%

Return on invested capital may be in the 10%–15% range.

8.3 Measuring Your Results

8.3.1 Achievements versus Planned Goals

The most important tool for measuring results is the business plan. The plan is what the management team will be trying to fulfill in the course of the fiscal year, and includes goals for sales, pretax earnings, and return on invested capital. Goals also frequently include EBITDA (earnings before interest, taxes, depreciation, and amortization). You may wish to add other measurements, but, as a rule, it is better to keep things simple and concentrate on basic matters. If the results achieved at the end of the year vary by more than 10% from the planned goals, examine the basis used for setting these goals. You do not want to consistently set goals that are either always surpassed or seldom achieved. A new team should be getting the hang of setting realistic goals within 2 years.

Progress should be monitored monthly, with in-depth reviews done quarterly. Do not be afraid to make adjustments, especially if a shift in the broad economy is clearly under way. Of course, making adjustments is not a substitute for finding out why the business is not hitting its targets and then fixing the problem; adjustments are appropriate only after you have done all you can to correct the situation and find that the cause is beyond the company's power to control. Keep projecting the plan when adjustments are made so that the next year's plan will have a baseline already in place when it is put together.

Some ambitious managers recommend setting "stretch goals," targets that go beyond those normally expected. However, the author strongly recommends that stretch goals should never replace regular goals, but only be used for secondary targets. There is some risk that your employees will not take other planned goals seriously if stretch goals cannot be met reasonably at least once every 3 years.

8.3.2 Financial Statements and Stock Valuation

These next set of tools are those that will be used by the board of directors, security analysts, and the investing public in deciding what kind of a job management is doing. The last group is the one that will confer its judgment via the price at which the stock of a publicly held company is traded. Quarterly and year-end financial statements should be fairly straightforward in showing how the company is fairing over a period of time, for at least two and often three years. If stock options are part of the compensation package, the price of the company's stock will be of more than passing interest to management. A depressed stock price may also be an attraction for corporate raiders. In the past decade, the stock market has been a harsh taskmaster, rewarding and punishing companies with dizzying swiftness. A lengthy period in which the company's stock price is depressed may also attract corporate raiders.

If your company's stock is trading at a multiple of earnings that is average or better for your industry segment, then all is well. However, if the stock price is tracking below average, you need to address what is wrong. Is it the company's performance or is it analysts' perceptions? If the latter, then it may be time to hold a press conference to tell the analysts how your company is doing better than average and something about its future prospects. Also, it may be that one or more large shareholders are selling off to diversify their portfolio; if this is the case, then the effect should be only temporary. You may wish to consider offering to buy back stock from large shareholders rather than having them sell to the public (some of whom may be speculators).

8.3.3 Customer Satisfaction

While you may be achieving plan, your financials look good, and your company's stock price is moving up, your customers must remain satisfied or everything will come crashing down sooner than you might think. Management needs to keep track of customer satisfaction, both formally and informally. How to do this? One way is to pay personal visits to major customers at least once a year. Talk frankly with them about whether or not your company is meeting their expectations and, if not, what can be done. This is not really anything more than what your account manager should be doing throughout the year, but your presence, as a senior or the top manager, is reassuring to a customer that their business is important to you and that you do not take them for granted. Such visits are an essential part of your job and ensure that you stay fully informed about how your customers view the marketplace.

Another, more comprehensive way, is to conduct formal benchmarking studies by questionnaires and telephone interviews. These latter studies are best done through outside, neutral parties, for example, consultants or market research firms, who have the necessary experience in conducting such surveys in a nonintrusive way. Mailing out bland survey forms to customers is not an acceptable or reliable way to learn how your customers view your company. In fact, such surveys are as likely to irritate your customers as they are to obtain genuine expressions of their concerns.

Customer satisfaction measurement is one of the most overlooked yet important tools that management has. Unless you know what your customers think of your company, you are effectively navigating by dead reckoning. Simply being nice to your customers is not enough; you must serve their needs, both present and emerging. You need to be doing such a good job at providing value with your products and services (much more so than your competitors) that your customers see your firm as an essential part of their supply chain. You also need to know *why* your customers see your firm's products and services of such value to ensure that if you need to make changes in your business model, they do not inadvertently harm your customer relationships.

Customer satisfaction studies are also a very important method of detecting coming changes in the marketplace, so that you are not caught by surprise. They can turn up important information about broader needs and problems (market intelligence) that you can use for directing R&D and strategic marketing projects. After all, your customers are experts on their business area; you would be hard put to find better, more credible sources of information about trends in their markets.

8.3.4 Competitive Rankings and Analysis

Every company should know how they compare to their competitors. After all, these are the people who are trying to take away the business you currently enjoy in the marketplace, and this is exactly what they will do if you misstep or lose your direction. Let us be clear that this is *not* advocating industrial espionage, which, by definition, is at least unethical if not illegal. Business intelligence, on the other hand, comes from many open sources of information which are not difficult to find if you know where to look. For example, a great deal of useful competitive information is available from the competitors themselves, in the form of their product brochures, annual reports, press releases, scientific papers, trade show exhibits, and websites.

The latter, in particular, frequently offers more information about companies' products, markets, plants locations and sizes, and personnel than any other single source. Sometimes, news releases and trade publication articles about your competitors will disclose production capacity and other information that is not otherwise easily found in open literature.

Competitors' customers and vendors are sometimes willing to share information, and these are often very reliable sources when it comes to estimating market share. However, it is usually more difficult to analyze and interpret this information than it is to obtain it. For this reason, it is strongly recommended that you use an experienced independent consultant to do competitive analyses, if they are to be anything more than financial comparisons (and they should be). Someone from outside your company often will be able to provide a more unbiased, broader view than your in-house personnel, when analyzing and interpreting the data.

A consultant will first seek your agreement on defining what really constitutes your competition, perhaps more broadly than the last time you considered this question. A nylon manufacturer will certainly consider other nylon producers as competitors, as well as producers of at least some other engineering polymers. Would you not consider zinc, aluminum, and magnesium producers to be competitors too? If you are a compounder, are your friendly glass suppliers not also competitors, if they are offering dry blends? If you are a custom molder or extruder, are not metal diecasters and metal extrusion companies also competitors? If you work with thermoplastics, what about those processors who work with thermosets? It is important to view your competitive environment on a broad basis to avoid being hit by one of those "unforeseen" changes mentioned in Chapter 1. You need some understanding of the strategies of these competitors so that you can both protect your business and exploit their weaknesses. It is instructive to learn how your indirect competitors view their businesses, as well. There have been times in the past when people in the plastics industry have been caught by surprise when trade associations representing glass, paper, or metal industries initiated a public relations campaign against plastics. One wonders why this should have been a surprise—after all, who is it that loses business when plastics displace conventional materials in packaging, construction, and other end uses? Construction unions have been an enemy of plastics wherever they see that converting applications from traditional materials, for example, metal or wood, to plastics, is likely to reduce their overall hours worked, threatening income if not jobs.

Next, your consultant will want your agreement on the structure of the study. You should want to know your competitors' identities, their size, growth trends, organization, product lines, major markets (and share in each of them), market share (both overall and by individual markets), innovation record, production capacity, and technical capabilities. Some of the information you want may simply be unavailable, such as production costs. Talking to people who are knowledgeable in the industry is more time-consuming and costs more than simply gleaning information from open sources, but such individuals can usually fill in missing details and, just as importantly, validate published data. This level of detail is normally unnecessary with respect to indirect competitors, but it can be useful to have some knowledge of who they are and what they offer. It is a good idea to validate competitors' production capacity as this information is sometimes "puffed up" just to discourage competition.

Do not ignore regulatory trends. Changes in government regulations often mean legal changes in what you can make, how you are allowed to make it, and how it may be used. Regulations can expand to add testing and reporting on some classes of products, as well as restrict or eliminate their use. This is an area that is always changing and requires ongoing monitoring. While your manufacturing and R&D personnel may be involved in these issues, this is an area where membership in an industry association can be very helpful in keeping you up to date and representing your interests before regulatory agencies and legislative bodies. The most stringent product regulations are generally found in Europe, followed by Japan, the United States, and Canada. Other countries tend to be less restrictive. Many companies, whose business is effectively global, aim to meet European standards as the simplest way to be compliant regardless of where their products are sold.

9

The Role of Acquisitions, Joint Ventures, and Divestitures

While every business tries to generate new technology, sales, and asset growth internally, there are limits as to how quickly or broadly this can be done. Acquisitions and joint ventures can achieve these goals more rapidly although they invariably bring management challenges with them. These problems are often serious; it has been said that over half of all acquisitions result in failure, in terms of meeting the original expectations.

Joint ventures are often a good way to become acquainted with an unfamiliar business before diving in all the way and making an acquisition; partnerships are also employed and are simply another legal form for joint ventures. One can acquire the other half of the joint venture or partnership later, if it is succeeding—or divest if it is not.

Divestitures are usually seen as an opportunity to redeploy financial assets by selling off business units that no longer fit with company objectives, as well as acquisitions that fail to work out. Sometimes, divestitures are necessary just to raise cash if the company is running heavy losses and cannot borrow enough to cover them. Almost every small, nonintegrated polymer manufacturing operation was divested many years ago because they simply could not compete against the economies of scale that larger, integrated polymer producers have. Today, very few people remember that Foster Grant (sunglasses) or Firestone (tires) or even LNP Engineering Plastics (compounding) had their own small, nonintegrated polymer plants as recently as the 1970s.

Acquisitions have played a prominent role in the evolution of the plastics industry. Many of these have been the result of the industry's small business pioneers being consolidated as the founding entrepreneurs died or wished to retire, but without having a successor. Also, some medium-sized firms found that they needed certain economies of scale in order to survive the competition from larger firms attracted to the growth and earnings potential of the industry.

Entering new markets or employing new technologies are other reasons for making acquisitions; thus, the emphasis is on the time saved by making such acquisitions versus the slower pace that internal growth and a learning curve otherwise required. Acquisitions are also a way to strengthen a company in a globalized marketplace in which large overseas firms may begin competing with your company.

Unfortunately, a significant number of companies available for acquisition are often in financial or operational trouble. These types of firms may offer a rare opportunity to acquire a fast and easy "fixer-upper" at a low cost—or they may turn out to be quicksand, absorbing far more personnel, time, and financial resources than were originally contemplated to effect a turnaround. A large—and bad—acquisition has effectively destroyed more than one company. These and other considerations are explored in the following sections.

9.1 Access to Markets

Probably the most common reason for acquiring another company or entering into a joint venture is to increase sales by broadening your existing market and customer base. While there are antitrust law limits on how far a company can go via this route, these usually only affect the actions of larger companies if the acquisition has the effect of reducing competition. Although the rules are somewhat elastic, one can usually count on attracting the interest of government antitrust attorneys if the combined companies will have an aggregate share of more than 10% in a readily definable and economically important market. Of course, everything depends on how broadly the specific "market" is defined. Fortunately, most plastics industry acquisitions can demonstrate sufficient competition from other materials that they should be able to pass this test fairly easily. For example, nylon 6 competes successfully with nylon 66 for a great many applications, so that a merger of two nylon 6 producers should be able to define the resulting position in a marketplace that includes both nylon 6 and 66, and not just nylon 6 alone.

Acquisitions that bring in one or more new—but related—product lines, or open up a new market for an existing product line, can make a great deal of sense. Economies of scale are achieved very quickly this way. For example, a specialized polycarbonate compounder might acquire a specialized nylon compounder and find that costs are reduced by using combined plant facilities. Combining sales and marketing, R&D, and administration may not yield smaller numbers, but will likely make it possible to handle growth more productively. Since these product lines are relatively distinct, there should be no antitrust problems attached to such an acquisition.

A favorite route to overseas markets is to acquire a local company in a target country; this can also be a high-risk strategy and is often more prudently pursued via a joint venture. The local partner provides customer contacts, knowledge of local conditions, and language abilities. The acquirer or joint venture partner provides technology and application knowledge, and, often, a source of supply. Ideally, both should contribute capital and management personnel. Overseas acquisitions require a great deal of patience, planning, and follow-up. Language barriers, differing business cultures, and more stringent government regulation are among the differences that one encounters vis-à-vis acquisitions in one's own country.

Stay away from making acquisitions in countries where bribes or "gifts" to government officials are commonly expected, or you will likely run afoul of the US Foreign Corrupt Practices Act. It is simply not worth the potential management time, legal costs, and exposure to bad publicity or even criminal charges connected with just the accusations that go with owning a business in such countries. Even if you do decide the risk can be managed, you may find that not paying bribes or making gifts can hurt your operations.

If you do proceed with a joint venture, and all or part of your manufacturing process involves closely held trade secrets, one may wish to move slowly and more securely by starting out with a joint venture that only *sells* products you export to the country for resale. After the relationship has become more comfortable and secure, then you can consider utilizing your trade secrets by manufacturing within the joint venture.

9.2 Access to Technology

Another important and sound reason for acquisitions or joint ventures is to gain access to a new technology, particularly if it is patented or not available under a reasonable license. It may be even more beneficial if the target company is on the leading edge of new technology development. Sometimes this is the solution to reaching sufficient critical mass of R&D assets and personnel to develop certain types of new technology. Other times, this is a way to avoid committing internally to a line of costly, high-risk R&D in order to discover certain types of new technology, although acquisitions are hardly risk-free.

Since relatively few small companies in the plastics industry have important technology assets, these types of acquisitions/mergers/joint ventures usually involve medium- to large-size firms and, therefore, significantly large amounts of money. This consideration, in turn, tends to tilt the table in favor of joint ventures as the most cost-effective way to obtain access to valuable technology.

Access to technology does not have to come via acquisition or joint venture, however. A properly constructed licensing agreement can do much the same with much less risk and capital outlay, assuming that the technology owner is willing to grant a license. Sometimes it makes sense to license relatively mature technology if your company wants to enter a field that is new to it; time savings and the avoidance of trial and error are often worth the license and/or royalty fees involved.

9.3 Manufacturing Capacity

Companies aspiring to increase and diversify their manufacturing base on a national or international basis sometimes find that acquisition or joint venture is the quickest and least costly way to do so. Companies that are ISO 9000 certified are desirable targets, because integrating them into the parent company will be simpler than bringing in companies that have less well-defined manufacturing practices. In a manufacturing acquisition, the due diligence phase should include a study to ensure that the target's production sites that duplicate the acquirer's existing ones are serving customers in geographic areas that cannot be reached economically otherwise. Possibly the new site or sites are so much more efficient that closing some of your old ones, or converting them to other uses, is part of the value to be obtained through the acquisition.

The most important asset is not the equipment, of course, but the personnel, their expertise, and proficiency. Machinery can be purchased and installed far more easily than people can be hired, trained, and become members of an effective team. The customers are also a very important asset, and they must be willing to continue to do business with the new owners or the acquisition target may not be worth pursuing.

9.4 Integrating Acquisitions into Existing Operations

Unless the acquired company was intended to be a stand-alone, independent operation, integrating the acquisition quickly into the acquiring company is absolutely critical to success. Reasonably detailed plans have to be drawn up and in place well before the acquisition is made, in order for quick integration to show any real results. Why integrate quickly? There are a number of reasons, and among them are the following:

- Motivating and retaining key personnel is vital to the success of the acquisition. You must give these people good reason to believe that they will be valued, and able to find at least the same satisfaction and opportunity in their jobs that they already have. If you do not, then they are likely to leave, or move in circles, if their direction or sense of importance has been lost in the course of completing the acquisition. People who leave often join competitors or even start their own competing business. This will leave you worse off than if you had never made the acquisition.

- Acquisitions are never inexpensive. You need to make these assets start earning a return for you as quickly as possible. "Transition time" is dead time.

- The planning process may reveal that the acquisition is not the right one, that it really does not fit after all. Better to learn this and abort the acquisition, than find out after it is too late. The cost of planning now will be far less than having to divest later.

- It takes time and planning to integrate ERP and other information technology (IT) and accounting systems. This simply cannot be done effectively overnight. The longer that the acquired company is out of your standard reporting loop, the less effective will be your knowledge and management of the new company.

9.5 When Not to Acquire

Knowing when *not* to undertake an acquisition is even more important than knowing when an acquisition is needed. Too many acquisitions are made for the wrong reasons, wasting huge amounts of money, time, and management focus that should have been spent on developing the business internally or finding another, more beneficial acquisition. Here are some of the more common "wrong" reasons for making an acquisition:

- Eliminate a competitor—This is almost always illegal and likely to involve the company in protracted litigation with federal or state governments. Litigation, especially antitrust litigation, is an enormous drain on management time and company funds that would have been better spent on improving the company internally. Moves such as this can also give a company a predatory reputation, which is not helpful to doing business. Furthermore, the existence of at least one competitor is actually a positive benefit—many potential users of your product are likely to be unwilling to expose themselves to the potential problems of depending on a single supply source.

- Wanting to show the industry how wealthy and powerful the acquiring company is—This calls into question why more attractive *internal* opportunities for more profitable growth have not been generated. Professor C. Northcote Parkinson in *Parkinson's Law* warned us that "building corporate monuments" is a sign of decline, not power. This can also be a sign that the board of directors is not independent, but subservient to a CEO who has an oversize ego.

- Diversification—Quite possibly one of the most common and very likely the weakest of reasons for acquisition or merger. Only rarely does management know enough about unrelated businesses that it can effectively oversee the operations of an acquisition that has little in common with your existing main line of business. Learning how a new business operates after the acquisition is much too late. And there is always the temptation for the acquirer to dictate how the business should be run, usually to the detriment of the acquired company. Furthermore, the stock market tends to punish the valuations of conglomerates that are not exceptionally well managed. Virtually every company that has acquired other businesses in order to "diversify" over the past half-century has wound up divesting them later—or being itself acquired and broken up. Still, this dismal track record does not seem to keep managers from acquiring unrelated businesses as a way to "smooth the economic cycle," or some equally attractive delusion.

- Acquiring a "bargain"—The urge to buy a troubled company on the cheap with the premise that it can be turned around quickly is usually a trap waiting for the unwary and inexperienced. Usually, it is very difficult to know from the outside what is the real cause of trouble or even if it can be fixed at all. The only certain thing is that troubled companies invariably soak up a great deal of management time that would otherwise go into improving your existing business.

- Almost as bad as making an acquisition for the wrong reasons is going about it the wrong way. This is a more subtle way to make an acquisition fail, but it seems to have an amazingly large number of eager practitioners. Since there are an almost infinite number of variations on this theme, let us just highlight a few of the real-life ones in the industry during the past several decades. The names are not revealed to "protect the guilty."

Changing the business focus from the one that made the acquired company successful to something new. History shows this blunder to be the favorite by far. It is not unreasonable to call it a blunder because it virtually never succeeds. One cannot help but wonder why the acquisition was made in the first place. One of the more egregious examples is the case of a commodity polymer producer that acquired an engineering plastics compounder for the purpose of making polyolefin compounds. The acquirer eventually divested the company when it came to realize that the equipment was unsuitable for the type of production it had envisaged. In the meantime, it had de-emphasized the engineering plastics business and lost many of the company's customers for these products. The new owner of the compounder was already in the engineering plastics compounding business and simply added this capacity and the remaining customers to its own—a far sounder basis for acquisition than "diversification."

A variation of this blunder is having the new acquisition report to a particularly ambitious and/or egotistical manager. Such a manager will never be able to resist the temptation to put his or her imprint on the new company, invariably changing the focus away from what made it successful to begin with. This may not be a bad plan if the acquired company requires a "work-out" from the verge of bankruptcy, but it is almost always a guaranteed plan for failure if the firm was a successful one. This scenario also holds if the acquiring company has a culture that demands its managers show "instant results" when they take on new responsibilities.

Replacing key managers in the acquired company with ones from the acquiring company. Unless the acquired company was a business failure, this makes no sense at all. Human capital is almost always the most important asset in any acquisition. If management does not respect the staff of the acquired company, then why did they acquire it? And, as noted earlier, those key managers have a way of leaving and then winding up in competition with the acquired company—and now with a grudge as well—not a pretty picture to contemplate and one that will hurt the acquisition in several ways. Noncompete agreements are not always upheld by the courts, so there is no foolproof way to keep key manager who have been dismissed from coming back to haunt you in the marketplace.

Merge several acquisitions together by creating a brand new top management group, while simultaneously reshuffling (or dismissing) the top management of the acquired companies. The new entity is expected not only to maintain sales and profitability of the acquired companies, but also to achieve higher revenue and earnings growth ("synergy")! This procedure does not appear to have any credible theory (or even any actual examples where it was successful) underlying it, but the lack of a reasonable rationale does not seem to keep people from trying it.

A large plastics conglomerate tried this by buying up regional distributors and trying to form one large national distributor from the group, dismissing or losing a number of key personnel in the process. The product lines of the regional distributors did not complement each other, and in some cases actually competed. This meant trying to persuade suppliers to agree to convert regional representations into national ones, a process that eventually had some success but only after a period of years. It also meant dropping or divesting some small but profitable representations that could not be converted. The net result was that the resulting mashed-together "super distributor" took years to equal the sum of the acquired parts—it is quite likely that the same net result could have been obtained—at less cost and more quickly—by acquiring one strong regional distributor and putting resources into internal growth and expansion.

This same company also tried a parallel strategy by combining a number of acquired compounders with dissimilar product lines. Next, some managers were dismissed, and the rest moved from their original positions (and competencies) to new positions. The "repotted" managers knew relatively little of the businesses or functions they now found themselves in. The result was a disastrous financial failure of such proportions that the parent company found it necessary to be acquired by another firm. The new owner had no choice but to dismantle most of the acquired structure, largely by closing plants, since much of the original business had been lost by then. The saddest part of this story is that many people in previously successful companies lost their jobs through no fault of their own while the incompetent executives who caused the debacle walked away with "golden handshakes." The better way to handle this process would have been to choose the largest and most successful acquisition as the "flagship" of the group and have the companies report to it. In time, the sales forces would be merged, as well as other duplicate groups. The process would have taken longer but would have had a far greater chance of success.

A particularly subtle way to cause an acquisition failure is to place the new company under a manager whose compensation depends on achieving certain goals that are not specifically compatible with the business form of the acquired company. As an example, a small, regional distributor was acquired by a large national distributor, and placed under an executive whose compensation is partly dependent on maintaining low working capital in those business units for whom he is responsible. Some of the most profitable elements of the distributor's business, however, depend on maintaining high stock levels for just-in-time deliveries to key customers. The executive (who had not been part of the acquisition team) was not persuaded that such customers could not be accommodated if stock levels were reduced by 10%. Not long afterwards, a last-minute order from one of the largest of the "last-minute" customers could not be filled because the new stock levels were too low to meet this need; the offer of a partial shipment was rejected. The customer then placed the order with another distributor who could meet the need immediately—and the customer then placed all future orders with the other firm. A better way to have avoided this problem would have been for the executive to request that his compensation goals be amended so as not to apply to except this particular situation.

9.6 Acquisitions versus Joint Ventures

As noted earlier, a joint venture is sometimes a superior way to achieve the certain ends, compared to an acquisition. If it turns out the joint venture was a mistake, it is often much easier and less costly to terminate it than it is to sell or liquidate an acquisition. Contrarily, a successful joint venture may offer an easier and less costly route to acquisition by buying out the partner's interest, compared to finding a similar but independent business to acquire.

Of course, a key part of any joint venture agreement is a previously agreed-upon procedure for termination.

Joint ventures are not without their disadvantages. Partnerships between widely dissimilar companies, in terms of size or interest, are seldom a wise choice. Relative dissimilar financial strengths or diversified interests tend to drive the partners apart rather than keep them together. The relationship between a mouse and an elephant is uneasy at best. The simplest joint ventures are normally the most successful, for example, a manufacturing joint venture where the partners split the costs and the output, and market the product independently of each other. R&D joint ventures are also relatively simple, but require a firm agreement on the partners' goals and their contributions of personnel and funding, as well as the intellectual property rights that will stem from any results.

The most complicated, and potentially contentious, joint ventures are where each partner is responsible for significantly different aspects, for example, one partner supplies raw materials and the other provides manufacturing (in its own facilities)—mutual agreement on transfer pricing and allocated costs is not easy and subject to frequent demands for renegotiation. Such arrangements can be better handled through a simple contract than by going the joint venture route.

9.7 Divestitures

Divestitures, the opposite of acquisitions, are thought to be necessary when an operation—or even a previous acquisition—turns out to be worth significantly less than expected or unsuitable for the company's purposes. In a real-life example, a polymer producer acquired a sheet manufacturer located in a large city. City building inspectors showed up soon after the acquisition closed and threatened to cite the company for numerous building and fire code violations unless they were paid bribes, a practice that had evidently been going on for years but had not been detected during the due diligence review. The acquisition agreement should have been deemed fraudulent and the transaction nullified. However, by then the former owners had disappeared, so the acquirer sold the business to a competitor that closed the existing plant and transferred the equipment, stocks, and some personnel to their principal location.

A less disastrous case might be that of a joint venture that has not achieved the goals set for it by one of the partners. That partner may then wish to divest its interests, most likely selling to the other partner. This is not an infrequent occurrence in joint ventures; in fact, it is a good reason to enter a joint venture as this is a much less costly way to explore a new business area that may not work out.

Another illustration would be that of a relatively independent part of the company or its business proves unable to show the required growth or profitability to keep up with the other parts. Mature product lines can fall into this category, particularly if they are, as stated, relatively independent of other lines but unable to support the company's overhead costs. Other producers of the same product may have an interest in acquiring such a business, or sometimes the managers of this business segment may wish to buy it and run it as an independent company. This is a good way to free up capital for investment in more rewarding operations.

Sometimes a capital-intensive part of a company's business has unmet capital needs that cannot be handled internally. This assumes that the business line is *not* a new, growing, and profitable one that might represent the future, because otherwise it would be better to divest lower growth lines in order to finance the more capital intensive one. A strain on capital also is a situation that might well benefit from joint venturing rather than outright divestiture.

Management may decide that the company must change its focus and a part of the company no longer fits with the new focus. If the part of the company to be divested is big enough, it may be "spun off" to the stockholders rather than sold. While this can be a legitimate basis for a divestiture, there are many cases where such divestitures have been less than successful for both the parent and the spin-off, so that all the ramifications must be considered very carefully before making such a move. As mentioned elsewhere, several major chemical companies in the last decade and a half decided they would focus on "life sciences" and sold or spun off some or all of their basic chemical operations. Neither parent nor stepchild company faired well afterwards; it appears that they may well have been better off to have kept the chemical businesses and spun off their life science operations instead. Profitable, growing businesses are not easy to find or develop internally. If they are contributing to the overall company and are not demanding otherwise scarce high-level resources, why divest? The better course may well be to retain them to provide cash flow for core and new businesses.

9.8 The Challenges of Being Acquired

Being the "acquired" rather than the acquirer offers special challenges. There are two basic situations to be considered: voluntary and involuntary. In the first case, for example, you have been seeking to find someone to acquire the small company you own, so that you can retire or pursue another career, or at least reduce your work hours and financial risk. In the second case, you may not have been informed or consulted in advance before another party acquired the company in which you are employed. In either situation, you may benefit from some guidelines about whether or not to stay.

9.8.1 Selling Your Company

Selling your company, especially if you are the founder, invariably involves a lot of emotion. Emotions often get in the way of making decisions based on logic. Under the circumstances, you will be generally better off if you employ a consultant to help you, someone who will point out where your interests are best served without the coloration of sentiment. You should contract with the consultant to pay for his time, not the value of the transaction. This will ensure that the consultant will not have a conflict of interest, if you choose to subordinate getting the best price to other considerations that are more important to you, such as continuing employment of your staff or keeping your facilities at their current location.

Although the dollar values are often small, the number of transactions involving plastics processors is much greater than in any other sector of the plastics industry. This is because there are so many small processors, most of which are owned by their entrepreneurial founders. The financial barriers to starting a processing company are relatively low—both equipment and building are often leased, rather than owned outright—just about anyone can go into business as a processor. As a result, the most important assets of a processor are the firm's customers plus its employees' experience and expertise. You will need to provide some assurances that these assets will remain with the firm after it has been acquired; having sales contracts and employee agreements in place are a convincing way to do this. Also, do not overlook the possibility that those same employees may wish to buy your business, conceivably through an employee stock purchase program. You will need expert legal and financial advice to set this up, of course.

Selling your company will be very time-consuming. Make sure that you have delegated as many of your more routine duties as feasible so as to have this time available as needed. If you intend to retire, ensure that you already have an effective successor in place that is ready to take over after you leave. Not many potential buyers will have much interest in taking over a small company if they must provide for managerial succession.

If you intend to stay on, you must accept the fact that you will no longer have the final say on how the business is run; that authority will reside in the new owner. It is common for a company founders to retire after not more than 3 years following sale of the business. This period will be spent smoothing the transition, particularly for key customers and employees. Even this brief time may be too long if you find you are having trouble letting go gracefully. Consequently, your contract should provide for an early release under such circumstances; you should expect to be bound by a noncompete agreement, following departure.

9.8.2 Surprise—Your Company Has Been Sold!

In the experience of the author, there is nothing quite as disconcerting as learning that your company is being sold. When such fundamental change is looming, one tends to imagine all sorts of potentially unpleasant consequences, making it hard to focus on day-to-day work. However, as the saying goes, "in change there is opportunity." If you are a senior executive of the company and the owners want you to participate in finding a buyer, then this situation can offer you the opportunity to develop important experience in the field of mergers and acquisitions (M&A). You will have the opportunity to demonstrate through your actions that *you* may well be the most compelling reason for a prospective buyer to acquire your company—and the best person to run it under the new ownership. Another possibility, if you have entrepreneurial talent would be to lead a management/employee buyout of the business from the current owners.

However, if the owners have not made you part of this process, do not become discouraged; things may yet work out. If your company is to be acquired by someone who wishes to diversify their holdings, there is less likelihood of management replacement or reorganization in your current company. On the other hand, if the pending company sale is the result of poor results, there is a strong likelihood that the new owners will make changes and it might not be a bad idea to update your resume. If the economy is in recession, your company is having financial difficulties, and the acquirer is intent on a consolidation, then your cooperation may be the difference between being kept employed, being dismissed together with others in a general reduction in force, or being given a severance package that is justified by just how willing and helpful you are in the course of the restructuring.

If you are not the senior executive in your company but rather someone lower down on the ladder, you have some thinking to do. Why was your company acquired? If it was for consolidation, is your business unit one that could be easily replaced by an outside organization? If you are in sales, R&D, or manufacturing, you will be asked to stay, more likely than not. Administrative functions typically bear the brunt of reorganizations. Also, employees in higher compensation brackets will be scrutinized carefully to see if they are contributing value in proportion to their pay. If you pass this inspection, look over the acquirer's operations to see if there is an attractive fit for you. Often, managers in companies that are being acquired for their growth potential are considered as good promotion prospects, and the new owners may see you as valuable to the company in another position. Always seek a positive approach to situations that are in flux; successful owners and high-level executives should want to retain such managers; if they do not, then you need to seek a position elsewhere.

10

Case Studies

To demonstrate how a number of the ideas discussed in the preceding chapters have been put into practice, this section will explore a few selected companies in the industry that, in the view of the author, have been managed exceptionally well over an extended period of time. Other firms might be equally successful, but the analysis of these particular firms is to make one or more points, not to conduct a contest. Furthermore, these profiles constitute only a "photograph" at a particular point in time and may well no longer be representative of their current condition when you, the reader, open this book.

10.1 BASF—Using Breadth of Product Line and Manufacturing Integration

BASF had the broadest, most complete line of polymers in the world 10 years ago but since then, has been selectively reducing its presence in products that have become true commodities, such as polyolefins and styrenic polymers (described in more detail later). It has been a key element of its business philosophy to be, as much as possible, a fully integrated polymer manufacturer, all way back to basic chemical and petrochemical feedstocks. Certainly, there are others, notably such oil companies as Exxon and SABIC, that also have fully integrated product lines, but these are largely limited to polyolefins. BASF has a German term for this integration, "Verbund," and says on its website (viewed 11/01/12), as follows:

> Our Verbund is one of BASF's assets when it comes to efficient use of resources. Production plants at large sites are closely interlinked, creating efficient value chains that extend from basic chemicals right through to high-value-added products such as coatings and crop protection agents. In addition, the by-products of one plant can be used as the starting materials of another. The system saves resources and energy, minimizes emissions, cuts logistics costs, and utilizes infrastructural synergies. Our global production Verbund is the foundation for BASF's competitiveness in all regions. With its six Verbund sites [two in Europe, two in the USA, one in China (a joint venture with SINOPEC), and one in Malaysia] and another 385

production sites, BASF supports customers and partners in almost every country in the world. The Verbund is all about intelligent interlinking of production plants, energy flows, logistics and infrastructure. Chemical processes consume less energy, produce higher product yields and conserve resources. BASF has long been recognized for making the most of its integrated approach to manufacturing, research and its overall management philosophy. This philosophy, together with the maximum integration of infrastructure, processes, energy and waste management, is known as "Verbund," a German word meaning "linked" or "integrated" to the maximum degree. Thus BASF's "Verbund" is a critical component of the company's sustainability program.

BASF was one of the first Western chemical companies to establish a presence in China. In addition to its Nanjing Verbund joint venture, it has more than half a dozen other wholly owned production facilities. In late 2012, it opened a new Greater China headquarters and an Asia Pacific Innovation Campus that will initially employ 450 people.

10.1.1 BASF's History in Plastics

BASF is one of the world's oldest chemical companies, dating back to 1865. It has also been, for a number of years, the world's largest chemical company. It has a proud history of discoveries; in the polymer area, its chemists are credited with the invention of nylon 6, polystyrene, SAN, ASA, and expanded polystyrene. Its engineers invented the gas phase process for polypropylene, among many other processes. BASF, long the dominant overall polymer producer in Europe, became competitive with other producers in North America in the 1980s, and began establishing an important and growing position in Asia during the 1990s.

BASF's emphasis on integration fits closely with its philosophy of making commodities profitably. When the profit goes out of a product, it either divests it or finds partners to increase the scale and share costs. Polyolefins are an example of commodity polymers that lost consistent profitability and were eventually moved into joint ventures of increasingly larger scale. BASF found that its initial scale of polypropylene production was too small to compete. Rather than simply adding internal capacity in this rapidly growing but extremely competitive product line, BASF increased its scale of operations by acquiring ICI's polypropylene and Hoechst's polyethylene businesses in the mid-1990s. When this strategy still did not yield improving profitability, BASF formed two joint ventures in 2001 with Shell Oil to pool their polyolefins businesses: Rheinische Olefin Werke to make ethylene and propylene via the cracking process, and Basell, to make and sell polypropylene and polyethylene. Unfortunately, this strategy did not prove to be the answer either: ROW was dissolved and Basell divested in 2005. Former

BASF executive and Basell Chairman Volker Trautz told this author in 2002, that, in his opinion, the boom-and-bust pricing fluctuations marking the polypropylene business, had "effectively destroyed" all of the investments made by producers of this polymer since its invention, so it was no real surprise that BASF decided to exit polyolefins. This business currently exists as LyondellBasell, an independent, publicly held company.

BASF was also a pioneer in styrenic polymers and acrylic resins, but went the joint venture route in 2011 with INEOS, in a new company called Styrolution, pooling their respective materials in a much larger-scale operation.

At the present time, BASF's polymer product line consists of nylons (including 6, 66, 610, and 6T), polyurethanes, POM, PBT, sulfone polymers, and specialty thermosetting polyesters. The only major thermoplastic that has never been in its portfolio is polycarbonate.

The company has experimented with forward integration into finished products, such as the production of magnetic recording tapes (which it invented in the early 1930s) and automobile components. However, it found these businesses to be increasingly less profitable than materials and also divested them in the 1990s.

In the 1980s, BASF even had its own distribution company, UltraPolymers, in the United Kingdom and Ireland. It entered this business by acquiring a bankrupt independent distributor that had been representing BASF. While UltraPolymers was a modest commercial success, it only served a relatively small geographic market and had no opportunity to expand this area due to other distribution agreements BASF had on the European continent. Consequently, BASF sold this business to Ravago, a large Belgian-based compounder-distributor.

10.1.2 The Effect of "Verbund" (Integration) on Product Line

As one might expect, the verbund concept has had a profound influence on the products BASF makes and its business philosophy. One of BASF's R&D targets is finding what materials can be made from the chemical intermediates it already produces. If a product shows good market potential but the monomers and/or polymers cannot be produced economically in-house, BASF has found a joint venture partner to attain the necessary economies of scale. As one example, BASF formed a joint manufacturing venture with General Electric Plastics to make PBT, at its plant in Schwartzheide, Germany. When GE Plastics was acquired by SABIC in 2007, it appeared that the partnership was no longer suitable for both parties, and BASF purchased SABIC's interests.

BASF's entry into acetal copolymer was originally undertaken as a joint venture with Degussa (now known as Evonik), in order to merge the technologies of the two companies to make a commercial product that was technically superior to competing products. Each partner had a manufacturing site and supplied raw materials, BASF at its main plant in Ludwigshafen, and

Degussa at its American plant in Theodore, Alabama. BASF was designated as the exclusive sales agent. In 2000, BASF acquired Degussa's interests in the joint venture, following Degussa's merger with Hüls.

BASF's strongest product presence is nylon 6, in which it is highly integrated. The company makes caprolactam from cyclohexane and then nylon 6 polymer, at its integrated sites in Ludwigshafen, Germany, and Freeport, Texas. The polymer is next finished and then either compounded or sold directly to processors, independent compounders, and fiber producers. At one time, BASF made its own nylon 6 fibers, but exited that business via a portfolio swap with Allied-Honeywell in 2003, that sent the fibers business to Allied-Honeywell and the engineering plastics business to BASF.

The company has by no means limited itself to high-volume materials, however, and has long been making a number of sulfone polymers for high-performance applications.

In summary, BASF has a long, successful history in creating and developing polymers that span the range of commodities, engineering types, and specialties. It has never been willing to accept diminishing earnings in a product line, however, and has usually fixed the problem by increasing scale, or divesting the product. This precept has led it to be the world's largest chemical company since 2006.

In late 2012, BASF announced a significant reorganization that split its plastics business in two, with engineering plastics and some specialty polyurethanes being combined with catalysts and coatings. Commodity plastics and the production of urethane monomers were moved to the chemicals unit. These changes were said to have been made to improve the company's ability to focus within its specialty and commodity businesses.

10.2 Victrex—A High Polymer Company

While it is not unheard of for a contemporary polymer producer to have a business based on one small family of high-performance materials, they are rare. A very successful example in this category is Victrex, which makes polyetheretherketone (PEEK) and a few close relatives, in the United Kingdom. ICI first began making PEEK in 1987 under the trademark Victrex. ICI decided to exit polymers altogether in 1993, and the Victrex product management group negotiated a buyout of the business it had been running. The polymer was protected by primary patents that did not expire until 2000, so that the new company was able to build a strong commercial business foundation without direct competition during this 7-year interval. Since that time, the Victrex team has developed a flourishing company that has consistently grown and stayed several steps ahead of the competition that has come into being after the basic patents expired. The company has integrated backward into monomer, regularly expanded production capacity, and integrated forward into making PEEK compounds, films, pipes, and coatings. Victrex has also developed some additional specialized polyetherarylketones, but the original PEEK is still their largest product. It has established a global presence with technical centers in the United States, Germany, China, and Japan to assist with the development of new applications.

As mentioned earlier, Victrex has seen competition come into being around the time its basic patents expired. In 1998, Gharda Chemical (India) introduced "Gafone," a competing version of PEEK; in 2001, Gharda sold this business to BP Amoco Advanced Polymers, which, at nearly the same time, was acquired by the large Belgian company, Solvay, for integration into its own line of specialty high-performance polymers. More recently, Evonik (the renamed Hüls-Degussa merged company mentioned earlier in the BASF review) has begun selling a Chinese-made PEEK ("Vestakeep") in both pellet and film form. Also, Arkema has recently added its own brand of PEEK ("Kepstan"—source unstated). While the heightened competition might seem worrisome, the emergence of additional large company suppliers of high-performance polymers is more likely to encourage the overall market for such materials to diversify and grow faster. Price is rarely a compelling reason to change suppliers of expensive, high-performance materials. In such fields as aerospace, the cost and time required to qualify an equivalent material from a new supplier is not easily justified.

Victrex "went public" in 1995, which has helped it to avoid debt to finance its expansions. Since then, the company has shown relatively consistent high rates of sales growth and profitability (approaching a 30% return on sales); it has paid a regularly increasing dividend. This extraordinary record makes Victrex as unique a company as the product it makes and markets.

10.3 LNP Engineering Plastics—Global Compounding

This case history was prominent in the first edition of *Strategic Management*. However, times have changed and LNP Engineering Plastics no longer exists as an independent compounder. In 2003, when LNP was acquired by GE Plastics (GEP) from Kawasaki, it was a global (ten locations), nonintegrated, specialties compounder, with approximately $285 M in annual sales, making it the largest such firm in the world at that time. Four years later, General Electric put its 75-year old GE Plastics business unit up for bids. Saudi Arabia Basic Industries Company was the winning bidder, renaming its new acquisition "SABIC Innovative Plastics." While its compounding plants are still in use, LNP has effectively become little more than a trademark in this huge new company.

Nevertheless, LNP has had a unique history in the plastics industry, and there are lessons to be learned from this history; hence, its inclusion in this section. It is instructive to see how each of its owners over the years treated LNP and what results ensued.

Since its founding in 1948, LNP followed a specialties business strategy that was very different from that of GEP. After reportedly paying over $310 M to buy this highly successful firm, GEP immediately replaced LNP's CEO with one of its own executives and changed LNP's basic business philosophy to conform to GE's own semi-commodity practices, for example, a dramatic increase in minimum order size and elimination of small-volume specialty products. It is noteworthy that within 2 years of the change in ownership, all of LNP's senior management team, as well as a number of key technical and marketing personnel, chose to retire or leave the company. Several ex-LNP personnel founded a compounding firm to service those small customers in which GE-LNP no longer showed any interest; it is healthy and growing, even in today's difficult economy.

From its founding in 1948 until its acquisition by GEP in 2002, LNP had worked its way up to become the world's largest independent engineering polymers compounder, with nine manufacturing sites located in North and South America, Europe, and Asia. LNP achieved this status despite being acquired and divested several times. Since the takeover of LNP by GEP and then SABIC, the title of "world's largest engineering polymers compounder" might go to PolyOne, which has acquired a number of smaller nonintegrated compounders in the United States and Europe (but has a large proportion of its business in polypropylene compounds). Another contender might be the privately held RTP Corporation, with a product line more directly comparable to the former LNP operation, and it has plants in the United States, Europe, Mexico, Singapore, and China. A third might be A. Schulman, which also has a global presence and a long history in compounding, although it started with such commodity polymers as PVC and polypropylene.

10.3.1 LNP's History

LNP was originally a privately owned custom cryogenic grinding company (LNP = *Liquid Nitrogen Processing*) when it was founded in 1948. In 1961, it adapted its process to making PTFE compounds and in 1964, began extrusion compounding a proprietary line of thermoplastic materials in Malvern, Pennsylvania. In 1970, LNP integrated backwards into making its own PTFE, nylon 66 and 610 polymers; unfortunately, the manufacturing scale was too small and the process technology too dated for the plan to succeed. The costs of building, operating, and maintaining the polymer plants resulted in large financial losses, nearly bankrupting the company. After a change of senior management, LNP sold off its polymer plants and returned to its basic business focus: compounding. This proved successful: the company became profitable again within a year and began to grow rapidly. In 1976, LNP was acquired by Beatrice Foods Company's Chemicals Division, where it continued to flourish. In 1985, Beatrice sold LNP to ICI, the British chemicals giant. ICI made LNP the North American regional center for its existing polymer business, but eventually decided to exit plastics altogether, selling LNP to a Japanese firm, Kawaski Steel, in 1991. Each of these successive owners left an imprint on LNP.

- Beatrice changed LNP management's focus on sales and earnings growth to one more oriented toward improving profit margins and return on invested capital. It also employed a decentralized management policy and was content to leave LNP alone as long as it regularly met its financial goals.

- ICI redirected LNP's North American marketing and technical primary efforts toward promoting ICI's existing line of polymers, which was underperforming in the United States. When ICI divested LNP, it kept the fluoropolymer compounds business and left LNP with only the thermoplastic compounds business. These moves first diffused LNP's focus, but then allowed it to refocus on its larger, higher growth rate business segment. In Europe, ICI more or less left LNP's operations alone, since ICI did not have the same need to grow polymer sales there as it did in North America. The result was that LNP's European operation retained more of the original entrepreneurial culture and focus on custom materials than the parent company did.

- Kawasaki brought ownership stability and patience, capital for expansion and acquisitions, and a return to relative business independence (Kawasaki's only other plastics activity was a thermoplastic sheet business in Japan). In early 2002, Kawasaki, feeling Japan's economic pinch, merged with another steel company. Wanting to concentrate on its core steel business, it sold LNP to General Electric Plastics. LNP's remarkable success while part of Kawasaki Steel, was bolstered by the parent company's recognition that it didn't "know" plastics, and so allowed LNP to operate virtually as a stand-alone company.

10.3.2 LNP's Business Strategies

While the author confesses he has a soft spot for a company he presided over at one time, his successor, CEO Bob Schulz, deserves great credit for LNP's business success during the near-quarter of a century he ran the firm, before it was "digested" by GE.

10.3.2.1 Focus on Customer Needs

LNP's primary successful strategy was simply to *focus on each customer's needs.* It faltered when it deviated from this strategy, such as the attempted backward integration step in 1970 and the fusion with ICI's polymer business, 1985–1991, and recovered when it went back to what worked so well before. It first began compounding customer-owned materials on a "toll" basis and then applied what it had learned about customer needs to making proprietary compounds. This approach also lead to making a large number of small-volume "niche" compounds, usually specific to unique applications. Before GE took over, LNP found that its typical order size stayed relatively steady at about one metric ton (2204 lbs); in other words, for every truckload customer, there were, on average, 20 single-pallet customers. Yet LNP's sales grew from $5 M in 1970 to $285 M in 2002, demonstrating that catering to small customers over a 32-year period can be a high-growth business.

10.3.2.2 Decentralize Manufacturing

LNP first expanded in 1964, by building a new facility in Thorndale, Pennsylvania, and moving its production there. LNP also added a plant in California in 1964, organizing the new location as a mirror image of the parent company in Pennsylvania, with duplicate functional management groups. This structure, while useful to give the new location a quick start, soon proved to be difficult to manage effectively; the various functional groups were later subordinated to those located at the home office. The next domestic manufacturing plant was built in Columbus, Ohio, in 1979; for purposes of order scheduling and logistics, it was run as an adjunct facility of the Thorndale plant.

In 1998, K-LNP (LNP's "parent" holding company, discussed later) acquired a polycarbonate recycling company, GHA Plastics (renamed RC Plastics), located in Houston, Texas. While this move gave LNP control over the quantity and quality of recycled polycarbonate feedstocks for its compounding plants, RC Plastics was operated as a separate division because RC's raw material sources, manufacturing technology, and retained external customer base differed significantly from those of the rest of LNP. RC's business grew and the plant was moved to a larger site in 2000.

LNP's European site was first built in 1968 in Breda, The Netherlands. This site was chosen because DuPont, LNP's primary PTFE supplier, had built a polymer plant in nearby Dordrecht and wanted compounds based on its Teflon resins (for its own resale) made close at hand. These facilities were outgrown soon after LNP added thermoplastics compounds, and the plant was relocated to Raamsdonksveer in 1976. The operation was run as an independent company with the functional groups coordinating their activities with the parent company groups. Soon after its divestiture by ICI, LNP acquired ICI's European "long glass" compounding plant, and relocated its Teeside, England plant to a new site in nearby Thornaby. In 1996, LNP acquired Eurostar, a French compounder with a plant near Paris, and merged it into the existing Netherlands-based European company.

LNP formed a sales office in Singapore in 1992 to develop business in burgeoning Asian markets and then followed with plant construction in Malaysia in 1995. While manufacturing and sales were run independently here, some technical and administrative services were being provided from the United States until the Asian company's business grew large enough to support local technical service.

In 1999, LNP acquired MIXCIM Indústria e Comércio Ltda., São Carlos, Brazil, as its first manufacturing presence in Latin America. LNP/MIXCIM was run as an independent subsidiary, with sales and R&D co-located at the manufacturing site.

In 2000, LNP announced the construction of a greenfield plant in San Luis Potosi, Mexico. This plant was managed within the North American Trade Association (NAFTA) as part of LNP's North American manufacturing operations.

Also in 2000, LNP announced formation of a marketing joint venture with Vetrotex America, the subsidiary of the French glass manufacturer, to sell "long glass" concentrates to injection molders in the NAFTA region.

It can be seen that LNP initially expanded at existing sites or greenfield facilities, but as it grew in size and geographic range, added acquisition as a selective tool to solidify its supply base and broaden its market reach.

10.3.2.3 Regional Management, Globally Coordinated

As LNP grew overseas, it did not attempt to direct these sites from the home office. K-LNP was set up to provide a central holding company to administer operations around the world in 1995. Figure 10.1 shows LNP's global structure under K-LNP. K-LNP's board, consisting of the division heads plus K-LNP executives, met quarterly to establish broad overall policy, review progress toward objectives, and consider whether any regional activities needed to be expanded further. Major customers were assigned to global account managers, who then coordinated meeting the customers' worldwide needs, regardless of where the manager was located.

FIGURE 10.1
LNP engineering plastics.

10.3.2.4 Patented Technology for Marketing Strength

LNP was unique among compounders in that it went beyond the usual trade secret approach to technology. It also sought patents on its new, advanced technologies and unhesitatingly enforced its patents against competitors. It had "inherited" three long glass compound patents from ICI, and decided to sue Ticona, DSM, and RTP for patent infringement in 1996. Ticona settled by paying royalties and cross-licensing its own patents to LNP; DSM elected to exit the business altogether. RTP refused to settle, and LNP took it to court; two years later, two of LNP's patents were ruled invalid but RTP was found to have infringed the third (however, the damages assessed were minimal). Overall, LNP scored two wins and a draw by aggressively enforcing its patent rights.

LNP also sought licenses from polymer producers to manufacture compounds based on unique materials, such as Dow's syndiotactic polystyrene, DuPont's amorphous LCPs, and Shell Chemicals' aliphatic polyketone "Carilon" (no longer in commercial production). These were not exclusive agreements, but they served to protect LNP's rights to develop patented compounds based on unique materials in cooperation with the polymer producer, but without fear of the results being shared with other compounders

In summary, LNP had a very simple but highly successful business strategy: concentrate on serving small, diversified customers with a full range of materials, including those specially created for the customer and the application.

10.4 Modified Plastics—Regional Compounding

Sometimes a niche market can be successfully defended against bigger, wealthier companies for an indefinite period of time. Modified Plastics in Anaheim, California, is good example. The relative geographic and time zone isolation of the American West Coast from the rest of the country makes it difficult to compete from outside the region against a skillful local firm.

Former LNP employees founded Modified Plastics in 1976 as a toll compounder, which has since expanded a full range of custom compounds. In 1990, Modified expanded by adding a color match and master batch business, Color Science. In 2008, a third company was added, Plastics Analytical Laboratory, to provide customers with testing and certifications. The combined companies—they are co-located—have grown steadily over the years, offering a full range of compounded products, and have become the largest West Coast compounder. Modified has further diversified its customer base to include those located in Mexico, southern Texas, Florida, and even Asia.

10.4.1 Using a Time Zone against Larger Competitors

In the 1980s, several larger Eastern or Midwestern-based compounders tried to displace Modified Plastics but failed. Each acquired or established a local compounding plant but then made the mistake of treating the plant as just another manufacturing site that ran on schedules set centrally. Plastics processors in California and the Pacific Northwest are numerous but typically smaller in size than in the other geographic regions of the country. They require fast service—often one-day—something that Modified Plastics does routinely and uses effectively as a competitive advantage. The geographic and time zone isolation of the West Coast mentioned earlier makes it challenging for companies in the American Midwest and East Coast to handle these requirements effectively from afar. Also, the firms mentioned earlier were not prepared to do business with a large number of small-volume niche product orders, preferring to think that customers would be willing to wait for these orders, if they were offered large-volume products at a better price. The problem with this strategy was that most small processors know from past experience that if they allow a supplier to "cherry-pick" their large volume needs, they will have to pay a stiff premium for their small-volume items—and certainly not receive overnight or even same-day deliveries. The net result is that the West Coast (and California, in particular) has remained a uniquely local market for plastics materials and parts. Furthermore, a number of West Coast end uses, such farm irrigation products, as well as computer and aerospace components, are not found widely elsewhere in the United States, giving this marketplace a something of a unique, self-contained character.

As mentioned earlier, LNP lost its 24-year-old fluoropolymer business when ICI sold the company to Kawasaki in 1991. This change had a particularly adverse effect on LNP's California plant, which was primarily a fluoropolymer compounding and reprocessing facility. While LNP tried to develop additional thermoplastic compound customers in Mexico and offshore, the plant's significantly lower throughput made profitability marginal. LNP eventually gave up and closed the plant in 2001, giving as its reason that its local customer base had emigrated to Asia. However, these were predominately large, global customers, not the small regional businesses served by Modified Plastics.

Modified Plastics now has a new challenge—the State of California has been raising taxes sharply and adopting onerous utility regulations that have greatly increased power costs, adversely affecting many manufacturers in the state. A number of California industrial companies have moved across the borders into Nevada, Arizona, and beyond, and more are planning to do so soon. Modified can service these firms from California, but not as easily as when they were based locally—and the financial pinch on Modified itself will need to be considered. Time will tell what happens next.

10.5 Maguire Products—Auxiliary Machinery

Located in Aston, Pennsylvania, this firm was founded in 1977, by entrepreneur Steve Maguire, who is still the owner and president. Mr. Maguire was recognized for his accomplishments by the Society of Plastics Engineers in 2012 through its annual Business Management Award. Maguire Products concentrates on simplicity, reliability, ease of operation, and value in the machinery it makes and sells worldwide. The company's major product lines include blenders, dryers, feeders, loaders, bulk box sweepers, purge recovery, liquid colorant pumps, extrusion controls, and access platforms. Unlike many larger companies, Maguire does not focus on maximizing sales growth as the primary goal, but rather seeking "slow, reasonable, and sustainable growth," a philosophy that has served Maguire particularly well during the Great Recession. Another unique aspect of Maguire's business philosophy is investing in additional manufacturing space to make operations simple and efficient: "because more space means less wasted motion, less movement of goods" (the company has eight buildings within a 2-mile radius of its headquarters). Inventory is kept at sufficient levels to ensure that production is smooth and uninterrupted, goods are always in stock, and customer orders are filled quickly. Purchasing and production are not driven by computer reports or projections, but rather by a "visual" inventory system that is simple to understand and implement, making it easy to decide when to order parts and when to build stock.

The owner's three sons are closely involved in the business, which should ensure continuity in the company's philosophy. The involvement of family members who work together as a team should also ensure that the company will remain independent for years to come, unlike many of Maguire's competitors who have been acquired and/or merged (some more than once!) in the past decade. The company's healthy, steady growth over the past 35 years is clear testimony to the success of the owner's business philosophy.

10.6 Common Threads

Just what are some of the more notable successful strategies that these different companies have in common in their successful quest for a growing, profitable business?

- *Adaptability*—All of these companies started with a basic idea for one or more products that were useful for certain end uses, but the idea did not end here. Each company was next able to broaden the applicability of the product to find additional end uses, either through product modification or by introducing the product to other end users whose performance requirements were similar to the first end use.

- *Integration*—Not every successful company has integrated manufacturing but it is very instructive to notice how many do. Integration helps contain costs as well as capture earnings that would have otherwise gone to suppliers. Integration also adds scale, an important consideration when the basic business is headed toward or has already taken on at least some commodity characteristics.

- *Dominant local presence plus global emphasis*—Small companies can succeed by being the best in a local geographic area, but when they wish to grow beyond a certain point, they must eventually go global. A regional presence is only an intermediate stage to going global.

- *Focused closely on customer needs*—If company management begins to concern itself more with internal matters than external ones (customer needs), the company is on its way to trouble. Each of the companies studied in this chapter are strongly focused on finding their customers, identifying their needs, and determining a profitable way to satisfy those needs. It pays to work with customers who have a "world view" in common with your own—this usually leads to faster and more profitable growth jointly. Never commit all or most of your assets to a single customer, though—diversification is essential to your company's survival.

11

Summary

What are the major lessons to be learned about the requirements of the plastics industry for successful management that differ from those of other industries?

- Successful management in the plastics industry is not the result of luck or the personal accomplishments of the chief executive officer. It is the result of applying the analysis, logic, and creativity characteristic of the scientific method used to discover technology, to the problems of planning, organizing, and executing business decisions, then inspiring and leading others to carry them out. The plastics industry is based on technologies, which advance constantly, and which shape the industry's markets and business models. These technologies, and their economics, are constantly evolving. These changes require management to continuously review company objectives and the means of accomplishing them to ensure that they are appropriate to ensure the company's continuing growth and profitability.

- Many other large industries, for example, construction, are not nearly as technology-focused as the plastics industry, and their technological evolution is much slower.

- The industry requires technically trained people to manage and run it optimally, but in an open and shared-decision style rather than by command-and-control. Upper management must ensure that the company's business objectives are current and clearly communicated to middle management on down. Again, we see that this differs significantly from many other large industries that do not require such a large proportion of their personnel to have such training, nor do they need such an open management style to succeed.

- The nature of the industry is for segments of it to be in frequent transition between different cultures of growth, size, and style. Timely recognition of these transition points and adapting to them is a top priority for management, because a mismatch of culture, style, and technology in a company will not produce successful results. Developing new products and applications requires an integrated technical marketing effort that is essential for success.

- The average life cycle of a product generally varies from very long to quite short as one moves downstream from polymer production to finished parts. This suggests that an increased focus on product and application development is required as one moves downstream. Note, however, that R&D intensity as a percent of sales is typically higher as one moves *upstream*. Companies that show consistent and above average growth and profitability are often downstream, because of their emphasis on developing new products and applications that allow value pricing. Value pricing is based on determining what your product or process is worth in use to the customer, rather than simply marking up costs to calculate a price. As a rule, it is easier for downstream companies to use value pricing than it is for upstream firms; for example, it is usually much more costly to make custom polymers or custom processing equipment, than it is to make custom compounds or custom parts. Value pricing, therefore, is an important consideration when reviewing your company's business structure, direction, and culture to improve growth and profitability.

- Acquisitions, mergers, and joint ventures are constantly reshaping various sectors of the plastics industry. However, one repeatedly observes that many acquisitions do not turn out well because they are undertaken for the wrong reasons or are executed badly, particularly with respect to acquired personnel. Sadly, this is not only a waste of scarce financial resources but also harms the lives and careers of the people in the acquired firms, through no fault of their own. It is a foolish waste of resources to treat the personnel of an acquired company as disposable unless contraction is an essential element to saving a business from bankruptcy and expansion is unlikely in the near future.

Index